Unfreezing the Arctic

Unfreezing the Arctic

Science, Colonialism, and the Transformation of Inuit Lands

ANDREW STUHL

The University of Chicago Press

CHICAGO AND LONDON

The University of Chicago Press, Chicago 60637
The University of Chicago Press, Ltd., London
© 2016 by The University of Chicago
All rights reserved. Published 2016.
Printed in the United States of America

25 24 23 22 21 20 19 18 17 16   1 2 3 4 5

ISBN-13: 978-0-226-41664-9 (cloth)
ISBN-13: 978-0-226-41678-6 (e-book)
DOI: 10.7208/chicago/9780226416786.001.0001

Publication of this book has been aided by a grant from the Bevington Fund.

Library of Congress Cataloging-in-Publication Data

Names: Stuhl, Andrew, author.
Title: Unfreezing the Arctic : science, colonialism, and the transformation
of Inuit lands / Andrew Stuhl.
Description: Chicago ; London : The University of Chicago Press, 2016. |
Includes bibliographical references and index.
Identifiers: LCCN 2016014741 | ISBN 9780226416649 (cloth : alk. paper) |
ISBN 9780226416786 (e-book)
Subjects: LCSH: Arctic regions—Discovery and exploration.
Classification: LCC G620.S894 2016 | DDC 971.9—dc23 LC record available at
http://lccn.locgov/2016014741

♾ This paper meets the requirements of ANSI/NISO z39.48-1992
(Permanence of Paper).

Contents

Note on Terminology

Throughout this book, I refer to the people living in what is today the western edge of the North American Arctic. I use first and last names wherever possible. Since the historical record usually lumps northerners into broad categories, I have made choices about terminology.

Since the indigenous rights and self-determination movements of the 1960s, Inuit in this region have referred to themselves with specific terms. "Inupiat" refers to the Inuit of Arctic Alaska affiliated with the seven villages of the North Slope Borough, the eleven villages of the Northwest Arctic Borough, and the sixteen villages of the Bering Straits Regional Corporation. "Inuvialuit" refers to the Inuit of the western Canadian Arctic affiliated with the six villages in the Inuvialuit Settlement Region. I use these terms when describing historical events from the 1960s onward. I sometimes add place names, such as "Inuvialuit residents of Inuvik" or "Inupiat of Point Hope." When alluding to all indigenous peoples of Alaska, including various Inuit and Northern Athabaskan communities, I use the term "Alaska Natives."

For episodes before 1960, I have chosen to identify indigenous communities with labels that are similarly historically specific, geographically precise, and culturally appropriate. For example, when discussing Aboriginal peoples who engaged commercial whalers near Herschel Island in the early 1900s, I may use the term "Inuit in the Beaufort Sea region." I prefer this name over "Canadian Inuit," since Inuit did not yet consider themselves Canadian and because these Inuit were not only living in what was then Canada. I deliberately avoid "Eskimo" in any case, because it is considered derogatory in Canada (even though it is accepted in Alaska). Above all, such naming conventions respect the principle that human relationships with the northern environment shifted over time. I would violate this principle by always employing "Inuvialuit" or "Inupiat" to refer to the Inuit of this corner of the continent.

Inuit is also a curiously plural term. It can stand with or without "the" as a modifier. One hardly ever sees or hears "the Inuits," at least in the places I have traveled for this research. If I refer to more than one Inuit person or more than one Inuit community, or Inuit communities from more than one political territory, I typically use "Inuit." The singular and adjectival version is "Inuk" (or "Inuvialuk" or "Inupiaq").

One of the trade-offs of my decisions on terminology is the litany of names that may be unfamiliar to the average reader. In coming to terms with these, I hope the reader also gathers a more sophisticated understanding of people and place—and the rich relationships between them.

Introduction: Is the Arctic out of Time?

Salt-and-pepper hair and a flannel shirt made Mark Serreze's style appear antiquated as he took the stage at the 2012 Inuit Studies Conference. His research could not have been more cutting edge, however. Using high-tech studies of the Arctic environment, the climate scientist revealed a startling vision of tomorrow.

Current conditions in the circumpolar basin, Serreze said, offer a window onto the future of a warming planet. With satellite imagery and computer models, he showed how greenhouse gas emissions from cities in North America and Europe drifted northward, concentrating above the Arctic Ocean. According to the US National Snow and Ice Data Center, where Serreze served as director, temperatures along the northern fringe of Alaska warmed twice as fast as the global average between 1979 and 2000. The scientist suggested that, in light of these phenomena, conventional relations of space and time would have to be revised. The polar ice pack—though thousands of miles from the industrial centers of the world—was no longer a distant location. Invisible pollutants and atmospheric circulation patterns directly connected north and south; the thawing of the Arctic was only the most visible effect of these ecological relationships. Perhaps even more alarming, Serreze cautioned, was that the shifting circumpolar landscape presented an early example of impending transformations for people and nature everywhere. To underscore these points, he titled his presentation, "The Arctic as Messenger of Global Climate Change."[1]

Stories like Serreze's pervade the media nowadays. Social scientists, journalists, and environmentalists also point to the northern rim as the "canary in the coal mine" of climate change. Like a fragile bird forewarning a grave danger for miners, melting ice has become a bellwether for many gathering storms. Economists envision shipping lanes in newly exposed waters—slicing travel time between Shanghai and Siberian natural gas and upsetting international economic orders. Policy analysts wonder if the tundra will become a battleground in a Third World War, as nations jockey for power over previously inaccessible resources. Investigative journalists foresee "the next Arabian Peninsula" unfolding when Big Oil collides with so-called ancient Inuit peoples. Greenpeace activists champion the protection of the region as "the most important fight in environmental history." In line with Mark Serreze's lecture, these onlookers capture the urgency of global change by characterizing a rupture between today's Arctic and the existing geological and historical records. One of the most common expressions they rely on—the "New North"—is also the most troubling.[2]

Suspend judgment about any one of these projections and consider this phrase. Like the larger arguments in which it is enrolled, the "New North" packs its punch not simply by compiling the facts of Arctic matters but by arranging them in relation to interpretations of the past. For its full effect, the "New North" requires two kinds of "Old Norths." The first is a remote and unchanging place, a wilderness that has been shielded from civilization until this very moment. The other is the reconstructed cryosphere, going back hundreds of thousands of years. Usually displayed on a graph, this "Old North" normalizes changes that occur over decades or centuries to demonstrate current Arctic temperatures as an anomaly of deep geological time. The tales of the "New North," then, are more than they seem. Beyond scientific models of the environment—or analyses of geopolitical conflict, natural resource conservation, and economic opportunity—they are histories. To portend the fortunes of global humanity, they first presume that the Arctic is *out of time*.

But is the Arctic out of time? In the pages that follow, I provide a straightforward response: No. Should that seem like a denial of climate change and globalization, I hasten to emphasize otherwise. If we want to respond attentively to intertwined social and environmental change in the Arctic, we need a history more nuanced than that of the "New North."

REPRESENTATIONS AND INTERVENTIONS, SCIENCE AND COLONIALISM

We start with scientists. Their concepts and research practices have accompanied efforts to conquer, cajole, civilize, capitalize, consume, and conserve the far north since the late 1800s. It may feel odd for some readers to link the pursuit of truths about the natural world with colonialism, especially in such a sparsely populated zone. But reflect for a moment on the most popular chapter of Arctic history: the search for the North Pole. As historians have deftly shown, American, Danish, and Norwegian explorers scoured the Arctic Ocean at the turn of the twentieth century not simply to seek out adventure but to find the world's remaining unclaimed lands. They returned south as national heroes because they expanded their home country's territorial possessions. Redrawing maps to display their discoveries, they sprinkled the names of sponsors on bays, islands, straits, and narrows. In these material ways, explorers extended Euro-American authority over the Arctic.[3]

Like their maps, explorers' tales about an Arctic terra incognita and savage Inuit helped southerners claim the north. With their words, they painted the Arctic as isolated and inhospitable. At times, they portrayed Inuit communities as backward for relying on Stone Age technologies for survival and primitive because they appeared to know neither religion nor law. In other instances, Inuit were strong—superhuman, even—for subsisting in an environment without trees, warmth, or readily accessible game. Historians analyze these depictions of the Arctic in relation to when they were produced and the dramas playing out at that time. Explorers' narratives were products of metropolitan societies concerned with the effects of overcivilization, a softness emerging from the urban existence that replaced life on the farm over the end of the nineteenth century. Those who consumed scientific ideas measured the Arctic against the norms of the Euro-American experience, reifying that experience as superior. An exotic place, one uninfluenced by machines and morality, then, was not an objective observation, but a view through imperial eyes.[4]

The ways historians have unpacked polar exploration illustrates the conceptual framework of this book. That is, scientific representations of Arctic nature are entangled with human interventions in the region. A map of the North Pole in 1909 was not an exact reproduction of the physical environment of coastlines and ocean currents. Instead, it was a document born from particular field experiences in particular places and widely held desires for global dominance. Philosopher Ian Hacking puts it this way: "Science is said to have two aims: theory and experiment. Theories try to say how the world is. Experiment and subsequent technology change the world. We represent and we intervene. We represent in order to intervene, and we intervene in light of our representations." Perhaps readers already understand this relationship intuitively, given the adage "knowledge is power."[5]

Polar exploration is just one example of the pursuit of knowledge converging with colonial ambition in the Arctic. Ideas of nature changed over time. New paradigms pushed researchers to ask different questions about the north just as human activity there altered the environment, requiring continued analysis. I use the term *scientist* to refer to an unwieldy group of people, from museum-based naturalists in the late 1800s to geophysicists working for military agencies during World War II to ecologists translating research into political activism in the 1960s. The differences among them are many, but what they share is this: at the moment they appear in this book, they compete for the mantle on definitive questions of scientific understanding, natural resource management, and national development. In these ways, the nineteenth-century specimen collector inhabits a social and intellectual space similar to the climate scientist today. By bringing these practitioners together in this way, we thread diverse representations and interventions in different historical periods in a line—thus reconnecting the Arctic's colonial past with its present.[6]

A TRANSNATIONAL ENVIRONMENTAL HISTORY OF SCIENCE

Like those crafting "New North" stories, I can hardly avoid the necessities of selection and emphasis that come with writing history. This study focuses on a corner of the circumpolar basin—a slice of North America above the Arctic Circle and between Kotzebue, Alaska, and the eastern edge of the Northwest Territories. Readers will not find a catalog of scientific inquiry about this region, but rather an analysis of how science served as a vector by which Canada and the United States each established colonial relationships there. I begin in 1881, when federal governments first sponsored scientific interventions in this part of the Arctic. I end in 1984, as federal officials recognized Inuit land title in binding agreements, in part because of the collaborative efforts of Inuit political organizers and activist-scientists. With these choices, I portray the far north in ways the media does not and other historians have yet to do. The seemingly unprecedented situation playing out across the top of the world today becomes the most recent of many attempts to understand, exploit, and protect Inuit lands.[7]

This is not to say historians have neglected the northwestern rim of the continent. There is a rich and sophisticated literature on Alaska and the Canadian north. This material, while by no means inaccurate, limits how we might come to terms with climate change and globalization, though. Northern historians have confined their analyses to national boundaries and hewed to divergent frontier traditions. If there is a grand narrative of Alaskan history, it fingers US economic elites as protagonists. Corporations and capitalists extracted wealth from nature—whales, gold, and oil, for example—despite the interests of local residents or the halfhearted support of the federal government. Canadian scholars have explained key social and environmental transformations rather differently. In their estimation, the north is a product of bureaucrats extending power—or failing to do so—by installing police and other security measures, creating an expansive and elaborate social services sector, and legislating conservation areas and hunting restrictions. While historians on either side of the border have presented northern colonial and environmental histories, then, they have straitjacketed interpretations of the modern Arctic. The region's connections to forces outside North America today may seem novel to historians concerned primarily with domestic relations. At the same time, the contrasting styles of narrating Arctic colonialism do not account for science, or prepare us to examine how scientific knowledge has tied together ecologies, economies, governments, and Native northerners over time. Indeed, scientists have remained peripheral in much of the existing literature.[8]

For northern history to serve as a frame of reference for our changing planet, it must be conceived in transnational and scientific perspectives. American fieldworkers preceded the United States and Canadian governments in the far north, which suggests the need to broaden the scope of historical investigations beyond the transfer or purchase of Arctic territories from the Russian and British empires. What eventually became understood as "Arctic" phenomena—in academic circles, or elsewhere—was the harvest of a network of researchers that spanned North America and Europe. Research on migratory animals, seminomadic peoples, and weather patterns also made clear that nature itself transgressed political borders. When federal governments—and, later, Aboriginal governments—put scientists in charge of "national" economic growth, these experts often relied on the infrastructure of multinational companies and the data sets of colleagues in neighboring nations. The Arctic is and was a global environment—just as colonialism has always been a project of knowing distant, different places. With gratitude, I draw heavily from northern scholars to situate science's history alongside political and economic expansion in Alaska and northern Canada.[9] When interpreting scientists' activities and interpolating them with accounts of environmental change, though, I turn to scholars elsewhere. As a result, the Arctic I portray may seem strange to northern historians, yet familiar to those curious about the interplay of capitalism, colonialism, and scientific exploration in other intemperate environments.[10]

For instance, what has been termed a cycle of booms and busts in northern resource extraction contains much more continuity when we follow the paper trail of scientists. In their published findings, researchers often ignored the economic or administrative activity in front of them to study plants, animals, ice, or Native communities. In unpublished works, like private journals or letters to friends, they let these comments run free. They detailed the operations of whalers, fur trappers, reindeer herders, and oil drillers, paying close attention to consequences for people and the land. Scientists could only get to the Arctic along transportation routes they did not themselves establish. Their travel logs thus become, in hindsight, documentation of the circuits of power that have linked the far north with corporate executives, bureaucrats, and consumers across North America and Europe. At the same time, Arctic physical conditions notoriously frustrated both fieldwork and industrialization. The forces of temperature, light, wind, and precipitation feature in scientists' reports, which confirms the ever-present yet multifaceted environmental factor of colonialism. In scrutinizing the scientific record in these ways, I connect successive waves of resource exploitation and exhaustion much in the ways environmental historians have done. Rather than see a frontier or a wilderness, we see a hybrid borderlands—one layered with human experience and its own ability to shape history. In this place, prior attempts at prying profit and knowledge from the Arctic—whether successful or not, by any standard—create the preconditions of subsequent interactions among people and nature.[11]

Of course, scientists do not tell the whole story of the Arctic. No history is complete. It is possible to turn this vice into a virtue, though. When we mark in time environmental changes and scientists' reactions to them—rather than ignore this past or dismiss it as anecdote—we gain a measure of humility. Humans make the world *as we know it*, even as the more-than-human world cannot be contained by science. This brand of history—a transnational environmental history of science—thus serves as a reference point for rapid global change today. It cleaves a beginning and an end from the hulk of experience, marks a pattern of cause and effect within it, and invites comparison between this pattern and what is happening now.[12]

Ironically, this is the appeal of "New North" narratives too. They make connections between the circumpolar region and activities of industrial society elsewhere. Their authors want nothing more than to raise awareness about the Arctic's relationship with the world. But just as they forge relationships among the far north, global ecology, and global commerce, they obscure the historical dimensions of global warming and globalization. It is tempting to think of the Arctic as a far-off land receiving the world's carbon emissions, and for this interaction to reflect the first collision of south and north, of industry and wilderness. It is richer to think of Arctic ecologies and cultures as having their own trajectories, which have been altered by patterns of consumption and circulation—for several centuries. This is not the first time the Arctic has been swept up into geopolitical debate, or the first time scientists have pronounced an environmental crisis there. What happens when we see the Arctic not as *just now* experiencing these events, but rather as experiencing them *yet again*?

DISRUPTIONS

After Mark Serreze left the podium, Nellie Cournoyea approached it. She wore a business suit, stylish glasses, and quiet confidence. Her keynote as chair of the Inuvialuit Regional Corporation provided a telling counterpoint to the climate scientist's interpretations of current conditions in the far north.[13]

For Cournoyea, that *Tan'ngit* (outsiders) could unleash profound change in the Arctic was not novel. It was a recurring theme for Inuvialuit—the Inuit of Canada's western Arctic. When whalers arrived at Herschel Island in the late 1800s, just offshore of Canada's Yukon Territory, so did their diseases, she said, wiping out entire villages and severing social bonds within those that remained. But Inuit repopulated; they did not go extinct. The money many made from assisting those in pursuit of the bowhead whale helped Inuvialuit reestablish themselves, and strengthen connections with Alaskan neighbors. In this example and many others, Cournoyea's Arctic was a story of heritage. If the Arctic appears to southerners as untouched, this is a function of their misunderstanding, as well as northerners' adaptations—not isolation. The northern landscape is and has been *resilient*, a term I use throughout this book. It names a colonial history of extraction—like emptying the Beaufort Sea of bowheads—and the capacity of people, plants, and animals to cope with the impacts of such exploitation.

When students of history consider indigenous vantage points, they can see Arctic colonialism more clearly. Science in particular loses its sheen of neutrality and objectivity. In Cournoyea's representation, for instance, Arctic research is much more grounded, more partial, more partisan even. Scientists come and go to specific locations. They arrive pre-programmed with hypotheses to test, which rarely answer questions of interest to locals.[14] And they are nearly always tailed by schemes of "modernization." Cournoyea told of Nuligak and Mangilaluk, Inuit leaders in Canada's Mackenzie Delta, who in 1921 rejected an offer of treaty by governmental agent O. S. Finnie. At the time, Canadian officials sought to formalize federal power over northern lands. Rather than money, Inuit wanted the government to help their poor and blind. But when Canadian agencies applied that money to introduce reindeer herding programs—a livelihood common in Siberia and Scandinavia, but foreign in Canada—Inuit in the delta found this culturally inappropriate and refused to participate. The government brought the herds anyway. They created a reindeer reserve where locals could hunt or trap only if they enrolled as apprentices under the supervision of white range managers, or if their application for a permit was approved. Inuit were never enslaved, forcibly conquered, or outnumbered by waves of settlers. Nevertheless, they were marginalized in their homelands, whether through the transformation of the physical environment that gave rise to their livelihoods or their exclusion from decision making over those lands.

Indigenous perspectives like Cournoyea's are crucial to understanding today's Arctic, because they disrupt commonly accepted notions of the region and implicate southerners in northern affairs beyond the fossil fuels they burn. Indigenous perspectives have not been well recorded by the pens of academic history, though.[15] For this reason, I have conducted research not only in archives across the southern United States and Canada, but in the Arctic as well. I spent twenty months living in Inuvik, a town in the Mackenzie Delta of Canada's Northwest Territories. I confess that my research in the field was not part of a grand plan hatched at my desk and then carried out in orderly fashion. I began by moving to Inuvik in the fall of 2007 as a community volunteer, with hopes of learning from residents about the impacts of climate change in the Arctic. I spoke with public figures, including scientists with local offices of federal agencies and representatives of municipal, territorial, and Inuvialuit governments. I heard many times that, while warming temperatures have altered physical conditions in the Mackenzie Delta, an understanding of climate change should be nested in the story of Arctic science and colonialism.

My interests piqued, I left for graduate school in 2008 to craft a more in-depth study. I returned to Inuvik two years later with a commitment to the kinds of community engagement recommended by environmental historians, historians of colonial science, public scholars, and the Inuvialuit Regional Corporation.[16] I attended town hall meetings and joined consultation sessions between oil companies and hunters and trappers committees. I sat in on northern studies classes at the high school. I pored over printed materials and audio recordings at local libraries, the Inuvialuit Cultural Resource Center, and the environmental agencies of the Inuvialuit and federal governments. I spent time on the land with Inuvialuit, soaking up stories of reindeer and trapping and hunting. When I found information about science's history in the Arctic, I shared it through formal presentations—to both town organizations and classrooms of high school students—and informal discussions with individual residents. I helped organize a five-day field trip that brought together scientists, Inuvialuit elders, and high schoolers to explore the culture, ecology, and history in Ivvavik National Park, on the border of Alaska and the Yukon Territory. Although I spent the majority of my time in Canada, I also traveled in Arctic Alaska. I completed a summer field course at the University of Alaska–Fairbanks, through which I visited North America's largest oil field in Prudhoe Bay and listened to residents of Anaktuvuk Pass talk about the historical nature of environmental challenges. Only when I headed south again in 2011 did I find that many of these practices accorded with the insights of postcolonial scholars, historical geographers, and Arctic anthropologists.[17]

Let me be clear: I do not wish to speak *for* Inuit. Inuvialuit and Inupiat speak on science, colonialism, and environmental change regularly, and have for some time.[18] My short stay in northern North America also did not yield complete and objective information. My personal background shaded my experiences, even as acquaintances, friends, mentors, and research participants helped me see different perspectives on the Arctic. I am aware that Inuit society transcends Inuit relations with scientists. Inuit have many more ideas about the Arctic than I present in these pages. I do not pretend to know these, or intend to discount them. Rather, I want to draw into bright context the ways scientific practices have configured life in the north—and vice versa. This history of science and colonialism in the Arctic cannot be fully told without learning from Inuit voices. Interviews, casual conversations, unpublished materials, and even silences in the historical record helped me apprehend the colonial nature of science, both then and now. Indeed, by confining studies to non-Native actors or neglecting indigenous accounts, historians contribute to amnesia about colonialism in the north and overlook the resilience of northern peoples and environments.

DISPLACEMENTS

Traces of history animate modern Arctic affairs and landscapes. Cournoyea shared the intersections of her own family with a genealogy of knowledge. Her grandfather, Mike Siberia, worked as a technician on a governmental survey of resources in the Yukon and Northwest Territories, the Canadian

Arctic Expedition of 1913–18. He married an Inuit woman who served as a seamstress for the scientists, sewing clothes from caribou so they could collect samples without succumbing to the cold. It was this expedition that gave Canadian officials the impetus for the reindeer project. Like scientists, members of Inuit communities played different roles over the episodes of intervention I examine. In the late 1800s, most were hunting guides. By the 1970s, many testified before public hearings on oil and gas development— like Cournoyea herself. This lineage shows that nothing in the Arctic is "pure"—not bloodlines, land, history, or science. I place the Arctic in time not to establish what belongs there, but to discern movements. Colonialism was, and continues to be, a tangle of intentional interference and its unintended consequences. As Cournoyea's testimony evinces, looking at a corner of the region, and keeping it in focus over the long view, helps apprehend both the course of history and the undercurrents of interruption—however fugitive or profound. Displacements in and from the Arctic, then, are part of the picture of global change I want to depict.[19]

For most of the nineteenth and early twentieth centuries, Inuit avoided outsider's attempts to control Arctic nature. They relied on knowledge and experience to live off abundant natural resources, including whales, seals, caribou, fish, and an array of plants. They also tapped into economies of exchange operating beyond governmental jurisdiction—whether through independent whaling or trading companies or fellow indigenous groups. Their lives, like those in the nonhuman world, hinged on mobility. By the mid-1950s, Cournoyea noted, Inuvialuit were stuck. They could no longer rely on the productivity of the land or the fur economies built upon it. At the same time, governmental scientists ceased to involve them. Many researchers no longer saw the value of local guides if they could complete an entire survey by plane while on the hunt for oil resources that played little part in indigenous society. Inuit stewed over these developments and began voicing their frustration—to scientists. The means of Inuit resilience by the 1960s, then, was not avoiding science, but appropriating it as a tool in political organization. Inuit allied with a group of concerned scientists—particularly those with environmentalist leanings—in their fight for sovereignty in Arctic Alaska in 1971 and on the Canadian side of the western Arctic between 1977 and 1984. In other words, as the military-industrial complex turned northward, and Arctic administrators granted scientists their highest levels of financial support and decision-making authority, a reconfiguration of science and power was set in motion.

Contrary to analyses of colonialism in the Global South, then, the end of my story is not the end of Arctic science. Rather, a century of intervention came to a close by articulating a science that was both humanist and environmental—and situating it carefully within civil society. When Inuit inked land claims agreements with the United States and Canada in the 1970s and 1980s, there was the fanfare of a revolution. But there was also a more subtle evolution, one predicated on the supervision of science and regular monitoring of the physical environment. I thus steer clear of "decolonization" as a reference to the period in which land claims were negotiated and formalized, since it implies a withdrawal of colonial forces. I use instead "postcolonial" to denote all of the years after 1881 as characterized by the legacies of the continued colonial encounter, with attention to the structures through which land was used, studied, and managed. These analytical choices affect more than disciplinary terrain. They comprise a way of being in the world that treats the past as constitutive of the present.[20]

In the 1970s and 1980s, both federal and Inuit governments emphasized the reorganization of knowledge production through more participatory and democratic decision making, informed consent procedures, and research licensing. Through these processes, Inuit shaped research agendas, resource development projects, and political discourse on environmental dilemmas at both the national and international levels. In all of these instances, Inuit forged closer—but no less contentious—relations with scientists.[21] It may be no surprise then, that Cournoyea closed her presentation before the Inuit Studies Conference by courting academics, cautiously. She expressed to members of the audience—many of them anthropologists, geographers, and ecologists—the Inuvialuit Regional Corporation's interests in collaborating on studies of sea ice and Arctic warming.

UNFREEZING THE ARCTIC

As the auditorium emptied, and conference-goers began to queue by the trays of cherry danishes, I introduced myself to the person in the aisle behind me. The requisite exchange of hellos and credentials ensued. But something in my introduction caught my new colleague off guard. "A historian?" he remarked, kinking his head. "What are you doing here?" In terms of academic standing, Arctic climate change is the domain of the natural and social sciences. It is not immediately obvious, to at least some of these scholars, what history offers to the study of environmental problems. The study of the past may seem like a hobby, a luxury in a bygone age of planetary stability. And, to be fair, many historians choose not to intervene in contemporary debates, even if their scholarship bears directly on present conditions.[22]

For those Arctic scholars in disciplines outside of history, this book provides a backstory for climate change and globalization. The Arctic has been central to the development of scientific knowledge since the middle of the 1800s, just as scientists have been central to developments in the Arctic over the same period. And, while melting ice is out of the ordinary, it is also nothing new: it is the latest in a string of rapid environmental transformations in the region over the last 150 years. When we take seriously these aspects of the Arctic's past, rising temperatures and shifting global power dynamics become more than ecological or social problems. They become matters of history as well.[23] Quite literally, previous episodes of scientific exploration and exploitation permeate northern societies and landscapes. Current frameworks of governance and knowledge-making in the Arctic are also products of the colonial experience, established through acts of Congress and modern treaties. If, in promoting ambitious plans for adaptation in the Arctic, this history is concealed, social and natural scientists may repeat it without realizing it. Geographer Emilie Cameron has convincingly shown that such politics are at play in climate science today, specifically in the literature on vulnerability and adaptation. Scientists in these fields often frame Inuit ecological knowledge and existing natural resource management regimes as inapplicable to physical conditions spawned by climate change. As a result, Inuit lose standing in channels of Arctic decision making. Natural scientists and social scientists can prevent this situation by examining circumpolar phenomena in the context of a history of colonialism there, and remaining committed to the principles of modern treaties and participatory democracy—like fair representation, informed consent, and self-determination.[24]

For my colleagues in history—and especially environmental history and the history of science—I intend these pages as a call to arms. We should engage more fully in current affairs, as public intellectuals with valuable and relevant expertise. As historians Mark Carey and Nancy Langston have argued for the issue of climate change, historians have an ability—and duty—to reframe environmental crisis in human terms. Circumpolar indigenous peoples have taken the lead on this, reminding world power brokers that climate change is not a problem of atmospheric chemistry—it is a human rights issue.[25] We historians with interests in Arctic and polar regions must complement this narrative by helping our colleagues in the natural and social sciences confront the colonial legacies of their disciplines. We must also challenge the pervasive misconceptions about the Arctic held in the communities we call home, misconceptions that, despite their deconstruction in the scholarly literature, continue to produce troubling political and material effects in the north. This kind of publicly engaged history is not meant to discredit scientists, or reject science in meeting environmental challenges. Rather, it reorients our discussions from the sites where warming is most obvious to the places and processes that drive climate change and globalization. In these ways and for these reasons, historians should intervene in civil discourse.[26]

The architecture of the narrative herein reflects this call to arms. Each chapter presents a case study of an environmental transformation in the western Arctic and its associated interventions. Taken together, they constitute a history of northern colonialism from 1881 to 1984. The chronology does not tally with standard narratives of the region, or even the neat numbers of the Roman calendar, but with the intersecting paths of North American politics, global economic history, Inuit accounts of the past, and science's internal development (disciplinary breakthroughs, for example). To foreground science as a prism onto all of these processes, I organize the chapters according to ideas that guided human interactions with nature at the time: dangerous, threatened, wild, strategic, and disturbed. Readers may find some of these terms anachronistic, unfairly confined to a period of years, or similar to the ways scientists imagine other "last frontiers" in Antarctica, Oceania, or the Amazon basin.[27] I welcome these reactions as part of continuing discussions of history and global change. The individual cases I present do not so much replace one another as entangle with each other, leaving today's landscape as the sum total of the past. This analytical and organizational approach stays honest to the historical records I have consulted, but it does something else equally as important. It underlines the historian's roles as interpreter and public scholar. That is, historians turn to the past not to discover what happened, but to make sense out of the world in which we now live.

As historian Howard Zinn observed, humans inflict pain on each other and the planet they share because they bury social atrocities "in a mass of other facts, as radioactive wastes are buried in containers in the earth."[28] The Arctic's colonial legacies are being buried, unwittingly, by a surge of popular and scientific media. Unfreezing the Arctic means that, as the circumpolar basin warms, we must dissolve notions of the place as frozen in time. It means confronting the long colonial life of an environment imagined as unspoiled, remote, or simply local—so that we can reach a more sophisticated understanding of climate change and globalization. It means releasing history from the past, showing how it matters now for all those contemplating the future.

CHAPTER ONE

Dangerous: In the Twilight of Empires

Imagine the village of Kittigaaryuit in 1861 (figure 1). There, on a beach at the mouth of the Mackenzie River, perched the largest known settlement of Inuit in the North American Arctic. Up to one thousand Kittigaaryungmiut—ancestors of today's Inuvialuit—sustained an elaborate community for half a millennium. They built semisubterranean houses out of driftwood and insulated them with the tundra's thick sod. During short but vibrant summers, hunters pooled labor and ingenuity to take beluga whales from the Beaufort Sea. They drove the animals into bays between the river and the ocean, using shallow draft kayaks designed to be easily carried, yet strong enough to withstand incessant waves. In the dark days of winter, they invented games and rituals—all meant to keep mind and muscles fit—and hammered out a system of governance through conversation and dispute resolution.[1]

To an untrained eye, Kittigaaryuit might look like highly localized adaptation. It could serve as an example of Inuit living according to what their immediate surroundings provide, however harsh and unforgiving. But when we look again, with a bird's-eye view, we see holes in that theory. Kittigaaryuit was one of a string of Inuit settlements draped across the rim of northwestern North America in the middle of the 1800s. Far from isolated or self-sufficient, these places depended upon an immense corridor of economic exchange, itself under threat and supported by two world superpowers—the Russian and British empires (figure 1). Across the Bering Strait, Russian American Company traders provided manufactured goods, like firearms and ammunition, to Siberian and Alaskan Inuit for fox, beaver, and sometimes reindeer. Their British counterparts, the Hudson's Bay

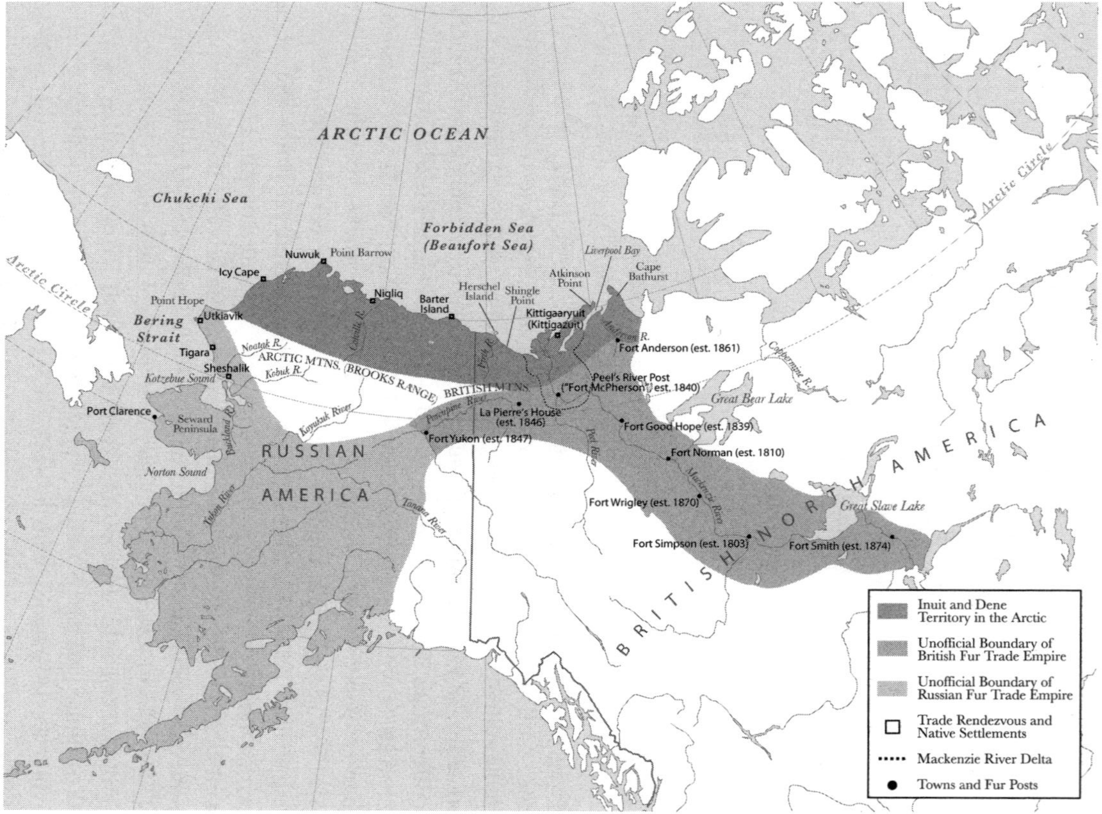

FIGURE 1. The western Arctic in 1861. Russian and British fur trading activities were consolidated around the Bering Strait and the Mackenzie River. Inuit living along the Beaufort Sea coast negotiated these empires and stemmed their northward expansion. Map by Morgan Jarocki.

Company, also desired access to Inuit country but could penetrate only as far north as Fort McPherson, several hundred miles south of Kittigaaryuit. At rendezvous points on Alaska's Arctic slope, Inuit from the Bering Strait region met Inuit from along the coast—from as far west as Kotzebue, Alaska, and as far east as Kittigaaryuit. Here, Russian-made products moved to coastal Inuit in exchange for seal oil and whalebone. Native northerners on the Beaufort Sea coast thus funneled the goods of two different empires toward their subsistence while preventing either from encroaching upon their territories.[2]

Now, hold on to this image of Kittigaaryuit as we fast forward to 1898. The place pales in comparison. The buildings remain, but they are shells of their former selves. Rarely occupied, their supporting beams have been scavenged to erect graveyards. Scores of Inuit fell victim to a series of devastating epidemics over the turn of the twentieth century. Inuit still move through this place, but they do not stay long. They are en route to another settlement called Herschel Island, just offshore from today's Yukon Territory. There they find the flow of goods and equipment that once sustained them, though a new imperial force has redirected and commanded it. With vessels that dwarfed Inuit kayaks, yet performed a similar function, American whalers sought the fat deposits and baleen of the bowhead whale. They transported these animal parts—themselves storehouses of nutrients accumulating in the Arctic ecosystem—to San Francisco for rendering into commodities that greased the machinery and pockets of urbanizing, industrial nations. Whalers helped convert marine life into illumination for factory workers, lubrication for power looms, and even fashionable corsets for ladies in Paris and London. In return, whalers brought their germs to the Arctic, as well as their flour and their rifles. Whalers traded all of these things with Inuit, transforming life on the Beaufort Sea coast.[3]

The American whaling industry triggered a sea change in Arctic history. Not simply because of the human and environmental displacements that followed in its wake, but because it set off a century of intervention, transformation, and scientific attention in the far north. To see how this could be so, we must examine movements within and across the Arctic at the close of the 1800s, like those witnessed in two scenes at Kittigaaryuit. Standard histories of this period often miss this action, focusing instead on political deal-making in Ottawa and Washington—think "Seward's Folly"—or the exploits of British naval officers in search of the Northwest Passage. While these events shaped the north, their influence came at a distance and mattered little to the future of the Arctic—according to Native northerners, at least. In Inuvialuit historical accounts, the coming of whalers, or *Tan'ngit* (outsiders), marks the shift between a traditional period (1300 to 1800) and the era of land claims agreements and climate change (1970s to now). That is, commercial whaling spawned colonialism in the north and thus provides the beginning of the backstory for modern global warming and globalization. In the view from the top of the world, British explorers and federal bureaucrats, those who achieved so much fame in Euro-American accounts of the past, left "little lasting impression" on the Arctic.[4]

As we follow scientific itinerants, we come to see the two versions of Kittigaaryuit in a different light. By the mid-1890s, the sun had set on Russian and British imperialism in northern North America, while the stars of the United States and Canada had risen. Whether in perpetual darkness or under the midnight sun the Arctic sat always in the twilight of empires.

THE SKINS AND SAMPLES OF AN EARLIER AGE

Furs were the "black gold" in a preindustrial era of global trade. The skins of beaver, muskrat, and fox linked consumers and fur frontiers across the continents. The opening of New World fur territories in the seventeenth and eighteenth centuries had tilted the global distribution of wealth and power toward North America. This shift accelerated in the nineteenth century, in part through the opening of western Arctic fur grounds, which boasted a stock of fur-bearing creatures unrivaled in Europe. In an effort to control the extraction and circulation of furs from Arctic land, Russian, British, and indigenous nations confronted one another over the end of the 1800s.[5]

The Hudson's Bay Company (hereafter HBC or the Company) wanted to tap the value of the Arctic and drain it via the Mackenzie River. The Company set up fur trading posts along more than 1,000 miles of the river by 1840. In that year, it opened the Peel River post, known today as Fort McPherson. Traders primarily engaged Dene First Nation communities nearby, but soon attracted Inuit living along the Yukon coast, in the Mackenzie Delta, and between the Mackenzie and Coppermine Rivers, to the east. The value of the Inuit trade grew over the 1850s, from 100 British pounds in 1854 to 1,000 pounds in 1858. In response, Company officers began to curry favor with Inuit, providing guns, ammunition, and tobacco at their request. These signs of encouragement convinced the chief factor of the Mackenzie District, James Anderson, to scout for a post situated squarely in Inuit territory. He called on Roderick MacFarlane, who journeyed to Anderson River in 1857 (figure 1).[6]

MacFarlane's experiences on the Anderson River provide a first reference point for the profound changes in the Arctic over the second half of the 1800s. A Scot who joined the Company as an apprentice clerk at nineteen years old, he earned the rank of manager after two years managing trade at others posts in the Mackenzie District. His efforts in opening up Inuit fur country in the 1860s linked him to a constellation of power in North America, one that shifted substantially with the arrival of commercial whaling to the Beaufort Sea by the start of the 1870s.[7]

The fate of the Anderson River post—the Company's first on the tundra in the western Arctic—suggests both the possibilities and obstacles of imperialism in the mid-1800s. After his initial reconnaissance of the Anderson River, MacFarlane wrote a glowing review of the fur resources of the country. The region, he concluded, boasted a large Inuit population, a suitable river for transportation, and abundant driftwood for fuel and shelter. After reviewing MacFarlane's recommendations, the North American director of the Hudson's Bay Company approved the construction of Fort Anderson in 1861. Over the next three years, MacFarlane reported increasing returns, with fox pelts bringing over 1,000 British pounds in 1864. By 1866, however, the Company shut the post down. Because the Anderson River was not part of the infrastructure-rich Mackenzie River valley, the Company struggled to administer it. An outbreak of distemper killed sixty-four sled dogs, which hampered already strained communication and transportation between Fort Anderson and the chain of posts that connected it to London. The end of the fort in 1866 thus foreshadows the combination of human and nonhuman forces that underpinned control over the Arctic for centuries to come.[8]

Despite MacFarlane's failures, or perhaps because of them, the Arctic earned a lasting place in circles of professional science. At the most remote location in the Company's system, MacFarlane had an unparalleled opportunity to collect materials from a landscape and culture not known by budding research communities. Between 1854 and 1866, MacFarlane contracted a scientific "fever," leveraging his seasonal interactions with Inuit and Dene trappers to collect specimens. In one five-year period at Fort Anderson in that span, he sent more than five thousand artifacts to the newly built Smithsonian Institution in Washington, DC.[9]

This alliance between MacFarlane and the Smithsonian was not uncommon in North America at the time. Like it supported science in the US West, the Smithsonian supported Arctic science with zeal, employing those already living in the far north as collectors. Its first curator, Spencer Baird, outfitted his Hudson's Bay Company team with dissecting kits, microscopes, pocket compasses, revolvers, opera glasses, and mosquito nets. In turn, Company men induced Inuit and Dene trappers to collect for them, providing American-made goods like handkerchiefs, jewelry, textiles, and the most prized item—double-barreled shotguns—in exchange for the skins of birds, bears, moose, and caribou. Baird also created user-friendly checklists for field-workers, like his "Directions for Collecting, Preserving, and Transporting Specimens of Natural History," so that his Company workers, regardless of their training, could contribute to science in an orderly fashion. Zoological collecting had long been conducted out of curiosity, with a fascination for the bizarre. In enlisting northerners with concrete guidelines, Baird sought to introduce an organized and systematic process of data collection and enfold the Anderson River—a place out of reach to the Company—into the grasp of the natural history museum. Between 1859 and 1867, more specimens left northern British North America destined for the Smithsonian than had ever left the territory for European museums.[10]

MacFarlane was not a prolific writer, but his collections in some ways speak for themselves. A recent report by a team of museum scholars and Inuvialuit resource specialists described its contents as including "a full range of skin clothing and bags; hunting and fishing gear; domestic tools, personal jewelry, and a wide range of pipes; carving tools and artworks; a model *umiat* (boat), sleds and kayaks." These objects were crafted locally but showed evidence of vast trade networks, which, for museum curators, now demonstrate "the vibrant, land-based lifestyle of the local people." According to Stephen Loring, who has reviewed letters between MacFarlane and Spencer Baird, Inuit in the region were well versed in trading relationships and used MacFarlane's presence to their advantage. "Already knowledgeable about the trading economy," Loring notes, "MacFarlane's Inuvialuit collectors took payment for the collections acquired on behalf of the Smithsonian in a wide array of non-Native commodities including clothing, tools, tobacco, and even guns." As they worked together to skin the tundra, it seems both scientists and Inuit took samples from each other.[11]

THE NATURE OF IMPERIAL FRUSTRATION: ANTAGONISTIC NATIVES AND ILLEGIBLE LANDS

MacFarlane's work deserves more interpretation, because even without a corpus to his name, he projected a powerful representation of the north. His Arctic natural history was a by-product of a mutually beneficial flow of material goods between Inuit trappers on one hand and British and Russian fur traders on the other. Yet this flow of goods induced fear in Russian and British fur companies as each vied for total control of Inuit lands, but failed to attain it. British audiences grew accustomed to stories of the dangerous Arctic. Its bone-chilling climate, extreme stretches of light and dark, and notoriously "barren" landscape had all vexed British naval officers in search of a Northwest Passage during the mid-1800s. MacFarlane's scant written material and sketches added another element to these treacherous conditions: in his mind, Arctic indigenes were equally as hostile as the physical environment.

In 1891, Roderick MacFarlane published a summary of his 1857 journey down the Anderson River in the *Canadian Record of Science*. He tempered his optimism about the natural resources of the Anderson River area with his interactions with Dene and Inuit. MacFarlane traveled with an Inuit "chief" between the lower stretch of the river and a larger camp of Inuit families stationed at its mouth in Liverpool Bay (figure 1). Here, a fleet of Inuit canoes and kayaks overtook his party, which, according to the British fur trader, immediately displayed their intent to engage in battle. "Seven guns were held up to intimate to us that they were as well armed as ourselves," MacFarlane wrote. He abandoned negotiations with these coastal Inuit, deciding to ditch his boats and walk seven days back to his camp upriver to avoid further confrontation. With similar expedience, he blamed this altercation on historic Inuit-Dene tensions around the fur trade. He did not acknowledge, however, that extending the Hudson's Bay Company's reach northward to the Arctic coast surely agitated Native northerners and their trading networks.[12]

This chronicle of events repeats a well-worn trope within Hudson's Bay Company lore about the Heroic Fur Trader and the Hostile Natives. In 1771, Samuel Hearne had traveled down the Coppermine River with Dene guides, hoping to initiate trade relations with the Inuit living near the Coronation Gulf. According to Hearne, his Dene guides ambushed a party of Inuit, murdering twenty men, women, and children. In 1799, Duncan Livingston attempted to replicate Alexander Mackenzie's 1789 journey down the Mackenzie River to the Arctic Ocean. After meeting a group of Inuit below Arctic Red River, Livingston's entire crew was apparently killed. Hearne and Livingston's experiences became legend, preserved in accounts that circulated throughout the Company for the next seven or eight decades. MacFarlane's representation of the western Arctic, then, fit into a well-defined set of representations and interventions under British imperialism in northern North America.[13]

FIGURE 2. Illustration of "Inuit chief" Noulloumallok-Innonarana from a sketch by Father Émile Petitot. Notice the knife in the right hand. Reprinted from E. Otto Hohn, *Among the Chiglit Eskimos by Father Emile Petitot: Translation of "Les Grands Esquimaux,"* Occasional Publication No. 10 (Edmonton: Boreal Institute for Northern Studies, 1981).

With the frame of these existing Company stories, MacFarlane's experience along the Anderson River encouraged him to promote the notion of Inuit as naturally antagonistic, aggressive, and inhospitable. To his Hudson's Bay Company superiors, he described Native northerners as tall and well-formed (figure 2). He worried about their "rather troublesome" character and their tendency to steal if not carefully supervised by white men. When corresponding with Émile Petitot, a missionary stationed in the Mackenzie

Delta during the 1860s, however, MacFarlane disclosed a deeper criticism of Inuit. Inuit were "sea bandits, who looked on theft, violence, deceit, unbridled libertinism, and even murder as virtues in which they took pride, while their womenfolk were shameless courtesans." Here was clear evidence of MacFarlane's views of Inuit, and how he tailored them for his imperial audience.[14]

MacFarlane's thoughts on Inuit were not contained to the page. They directly influenced the architecture of empire in the north. Take the construction of Fort Anderson. The fur post was the physical embodiment of outside intervention, a space wherein British traders extracted profits from the Arctic and displaced its natural resources. Because MacFarlane and many of his associates feared Inuit, the Company built Fort Anderson to allow careful surveillance of all activities inside and outside its walls. According one British trader's description, the site of trade negotiations was "surrounded by a square palisade about fifty-five yards long with a twenty-foot-high bastion at each corner." Given that the Arctic did not allow for easy shipping—and provided few trees—the lengths to which the Company sought to secure Fort Anderson are telling. They speak not only to the value of mercantilism but also to the perception of danger that accompanied imperial exchange in the Arctic.[15]

In most analyses of colonialism, scientific representations foster exploitation because they allow administrators in distant locations to read the landscape according to colonial desires. In other words, scientific stories and images make a foreign and confusing space legible—they rationalize nature for the purposes of resource extraction. MacFarlane's Arctic reflects a different kind of imperial experience, an exception to what became the rule in colonialism's next century. The dangerous Arctic indicates illegibility, the inability to bring the region into the orbit of British power. MacFarlane's science nurtured an aching frustration that Inuit and the Arctic environment would always get the better of the Hudson's Bay Company.[16]

Such a conclusion comes into sharper relief when we consider Russian and Inuit roles in the fur trade. By the time Fort Anderson opened in 1861, the influence of the Russian empire could be felt far to the east, among the Inuit groups along the Beaufort Sea coast. In 1865, missionary Émile Petitot visited the Hudson's Bay Company's Peel River post and witnessed the arrival of Inuit from the Colville River Delta in Alaska (figure 1). He noticed in their party a mixed-race Native with red hair and freckles, probably of Russian descent. Petitot thus had firsthand evidence that commerce in the Bering Strait had ramifications across Alaska, through Inuit territory, and into the domain of the Hudson's Bay Company. That Petitot recorded his encounter with this Inuit man was not merely a matter of fascination—it was also economic strategy. Many Inuit along the Beaufort Sea coast did not live in the vicinity of a Russian or British fur trading post. To Russian and British fur companies, this area represented a frontier, one that would either extend the Bering Strait trade farther into the New World or allow the Hudson's Bay Company to finally lay claim to the last, possibly most lucrative, corner of the continent.[17]

On the other hand, in an Inuit perspective, the Beaufort Sea coast constituted a homeland, which could be protected through engagement with the fur trade. Inuit developed their own extensive trading networks spanning the region from the Bering Strait to the Mackenzie and Porcupine River districts. Their rendezvous occurred in the summer at locations along the Arctic coast in Alaska and Canada, like Nigliq, Pigliq, Barter Island, Sheshalik, and Kittigaaryuit (figure 1). Trading fairs helped Inuit manage the power of the Russian American Company and the Hudson's Bay Company. When tensions with Dene at the Peel River post ran high—threatening to deprive Inuit of goods, equipment, and food—Inuit could rely more heavily on the Bering Strait trade. Because Inuit along the Mackenzie Delta and Beaufort Sea coast did not immediately embrace the Hudson's Bay Company, British officials presumed that the Russian American Company had made significant inroads into British North America. They could not imagine the kind of trading networks developed by Native northerners. Underlining the impulse to expand British control over the Arctic and curtail Russian influence, one Company director identified the mission of Fort Anderson as "intercept[ing] the trade carried on by the Esquimaux with the Russian territory, with which they have been in the habit of carrying on their traffic." At least until the arrival of American whalers, then, Inuit situated between the Bering Strait and Mackenzie River were able to negotiate the terms of the trade between Russian and British empires—even if they could not control how scientists imagined them.[18]

FROM FUR COUNTRY TO WHALING FRONTIER: THE *TAN'NGIT* ARRIVE

When the United States and Canada acquired northern territories in the late nineteenth century, they inherited a place already connected to global markets and global empires. The importance of the fur trade was not lost on the American Commercial Company. In 1869, Captain Raymond of the

US Army traveled to Fort Yukon, a Hudson's Bay Company post at the top of the Yukon River, to evict the Russian American Company. Despite their continued operation of the fur trade, US political, commercial, and military leaders redirected their energies toward the northern seas. In the minds of many bureaucrats and boosters, Alaska held a strategic location in the global economy because of its access to the "Great Circle" route of the North Pacific and the Asian markets within it. The purchase of Alaska in 1867 drew criticism, with some critics infamously dubbing the move "Seward's Folly." Yet many others supported it, including those in the American press, on Capitol Hill, and in commercial centers on both coasts. Recognizing Alaskan assets was one issue, but capitalizing upon them was another. The nation's grazing lands, mines, cities, and farms could not easily be linked to Alaska by road or rail. The colonization of the territory thus commenced by water-based transportation.[19]

After purchase, Congress enacted a period of military rule in the territory until 1877, to protect and expand its reach and that of the fisheries industries in Alaska. In addition to the booming salmon canneries on the Yukon River and the seal harvest in the Pribilof Islands, the government supported the commercial whaling industry, which gradually explored the northwestern Alaskan coast and Chukchi Sea in search of bowhead whales. In the three decades leading up to 1900, whalers in the Arctic harvested thousands of bowhead whales, earning some of the largest profits in American whaling history. The Navy, the Army, the Signal Service, the Revenue Cutter Service, and the U.S. Coast and Geodetic Survey all sent representatives on reconnaissance missions in Arctic Alaska in support of whaling during this time.[20]

A suite of factors helped make the Beaufort Sea a site in the far-flung geography of American empire. As historian John Bockstoce has demonstrated, depleted stocks in the Pacific Ocean and Bering Strait provided an initial incentive for whalers to press the boundaries of their harvesting areas. After hearing reports of abundant whales in the Beaufort Sea from British Royal Navy officials in the 1850s, whalers flocked there. Foraging crews engaged Inuit living along the Alaskan coast north of the Seward Peninsula (figure 1). By 1879, Captain E. F. Nye could write to the *New Bedford Standard*, a prominent journal within whaling communities, describing Point Hope Inuit as knowledgeable traders, who recognized "the value of whalebone and skins." If overharvesting pushed whalers north, the ecological riches of the Beaufort Sea pulled them there. The sea receives the flow of several northern Alaskan rivers and the Mackenzie River, Canada's longest inland waterway. Nutrients wash off the land toward the shallow waters beyond the coastline. There, they provide essential ingredients for phytoplankton, the floating plants upon which bowhead whales feed. Historical ecologists estimate the population of bowhead whales prior to the American whaling industry to number around thirty thousand individuals.[21]

The bowhead appeared as a species uniquely suited to the needs of the whaling industry and vibrant global markets. Whale products featured prominently in metropolitan culture throughout the second half of the nineteenth century. Whale oil found its way to lighthouses, candle makers, and factory machines, while baleen formed corset stays and buggy whips. The average bowhead carried more oil and more baleen than any other whale sought by the industry. Once extracted from the Arctic, economic machinery in San Francisco processed the animals. Commercial men established elaborate whaling companies, purchased boats from builders in New England, hired seasoned captains from around the country, and employed "land sharks" to find crew members in the Bay Area desperate for work. The most prominent of these companies, the Pacific Steam Whaling Company, had its headquarters in San Francisco, where it consolidated the hauls from its various ships and distributed these to manufacturers in New York, London, and Paris via ocean and rail lines.[22]

With the benefit of hindsight, we can see these factors converging and thus creating the possibility for an Arctic whaling industry on the Beaufort Sea coast in the late 1800s. But in written accounts by whalers and other northern travelers in the 1870s and 1880s, American empire's northward expansion was far from inevitable. For much of the century, the dangers of ice thwarted many efforts to reach the "East Shore" of Alaska—that coast around the northwestern hook of the territory. As whaling vessels could harvest only in open seas, whalers timed their entrance and exit so as to allow for the greatest potential catch and a safe passage home. Ice retreated from Icy Cape around June of each year and returned to freeze-in any remaining ships by late September. When depleted stocks forced whalers to test the limits of this open-water season, whalers often risked their life for their livelihood. In 1871, ice crushed thirty-four whaling ships off the coast of Point Belcher. In 1876, the freezing seas consumed a dozen more. Even as steam power helped alleviate many of these issues, ice remained a force that held in check the economic appetite and geographical expansion of the whaling industry. Indeed, for much of the late nineteenth century, the Beaufort Sea had no other name but "Forbidden Sea" (figure 1).[23]

By 1880, influential investors in New Bedford (Massachusetts) and San Francisco—the powerhouses of the American whaling industry—lobbied the federal government to protect and support the extraction of bowheads. Lobbyists promoted the industry as integral to the nation's economy, as it employed more than 1,500 men each year and represented a capital investment of more than $1.4 million. Industry representatives urged the government to erect life-saving stations on land to assist whalers who escaped from wrecked vessels. Ship captains noted that the normal course of responding to disaster—to "escape by small boats to the shore" and proceed home through communication with local settlements—would be impossible in the "uninhabited and desolate regions." The US Revenue Cutter Service eventually met these requests. The service, which had patrolled the waters of the Bering Strait after the Alaska purchase, gradually explored the northern coast of Alaska until the early 1880s. Revenue Cutter Service staff confirmed the viability of a shore-based relief station on the Beaufort Sea at Point Barrow, Alaska, in 1884, after careful exploration during the International Polar Year of 1881–83. This event—the landing of the whaling industry at Point Barrow following a government-sponsored scientific venture—signaled the dawn of another century of colonialism in the Arctic.[24]

SCIENCE AND EMPIRE ON ICE

The century of science, intervention, and environmental transformation at the center of this book properly begins with the International Polar Year of 1881–83. We have already seen that the age of British and Russian imperialism in the Arctic incorporated scientists and generated landscape change. We can set this year apart from that earlier era, though, for a few reasons. First, the Polar Year earned sponsorship from US federal agencies—the national museum and the Revenue Cutter Service—to deploy trained scientists to collect knowledge in direct support of resource extraction, without regard for issues of sovereignty or ownership. Second, and more importantly, the enrollment of professional science in service of US empire displaced Inuit in ways imperial fur traders never did. Whalers and naturalists established themselves as experts over Arctic nature, developing a body of knowledge about marine life, surf, skies, tides, and ice that rivaled that of Inuit. Fur traders and collectors had relied solely on Inuit as harvesters, interacting with them mostly at the post. Now imperial agents used science to effect exploitation within the same territories where Inuit operated—a direct challenge to Inuit authority.

The details of the International Polar Year draw out these distinctions. If science and commerce had been related in British and Russian empire, they were married in the American version. Under the command of Lieutenant Patrick Ray, ten men left from San Francisco for the Alaskan Arctic slope with the mission of making "a three-year unbroken series of weather and magnetic observations." Their task reflected both the importance of Arctic whaling to the US economy and the interests of world scientific communities. At the close of the 1870s, scientists convened a series of international conferences, wherein the United States joined Austria, Norway, Sweden, Denmark, and Russia to coordinate global observations of meteorological phenomena in the circumpolar basin. At the time, meteorologists understood that "the weather in high latitudes held the key to the behaviour of the atmosphere on wide scales." Perhaps not surprisingly, weather and navigation studies were also "of practical interest to the whaling industry," which had lost so many vessels in the Bering and Beaufort Seas in the 1860s and 1870s.[25]

The scientists recruited to carry out these duties were unique from earlier collectors in Arctic history, even if they came from a familiar institution. The Smithsonian's two recruits for the International Polar Year, naturalists John Murdoch and Middleton Smith, held discipline-specific training. They were not, in contrast, fur traders pulling double duty as gentleman collectors, as was the case with Roderick MacFarlane in the early 1860s. Murdoch held a master's degree from Harvard in zoology. When the Smithsonian invited him to go to Alaska, he was teaching zoology at the University of Wisconsin–Madison. After returning from the north, he worked with leading botanist Asa Gray and took a position as librarian at the Smithsonian, where he consulted the expanding Arctic collections in crafting a series of publications about Inuit culture in the 1890s. In ways similar to Murdoch, scientists in the coming century leveraged the expansion of the North American political economy for the elaboration of professional careers in science.[26]

As testament to the role of the Beaufort Sea in academic and colonial development in the United States, whalers and scientists developed a comprehensive knowledge about ice. The writings of Hartson Bodfish, a New England–born whaler who amassed more than thirty years of Arctic experience, suggest the success of a whaler stemmed from their hydrological education. Bodfish offered vivid details to readers, surely to help them avoid crushing defeat in an ice pack. When about to freeze, the ice would "mass together and keep getting larger and larger, then it piles into points and ridges," he wrote. To determine if existing ice is new (and thin) or old (and thus a threat to a ship's hull), take note of its color, he said. "When the ice is new it is slightly darker than the water," he cautioned, "but most of it is dirty from the dust blowing from the shore and looks more like mud than anything else." Whalers recognized relationships between ice and wind, as changing wind directions and intensities could shift the ice pack instantaneously, presenting hazards to vessels. Knowledge about ice and wind also helped whalers locate whales. In explaining the migration patterns of bowheads in the Alaskan Arctic, whaler Jim Allen noted, "I would say it's the ice and wind conditions that make the difference. There are no two years the same in the Arctic Ocean."[27]

Scientific fieldwork also concentrated on the interface of surf and sky. In doing so, scientists reoriented imperial knowledge from terrestrial ecosystems and economies in the mid-1800s to their marine counterparts by the 1890s. Because the expedition's primary objective was meteorological, Murdoch and Smith were required to serve "two four-hour watches each day," during which they "recorded general weather conditions and took hourly meteorological readings." Smith's journal entries reveal how this position trained him to look outward and upward—and not at the ground in front or behind him. He recorded the daily levels of barometric pressure, temperature, prevailing wind direction, and wind velocity, eventually computing monthly averages for these measurements. He also noted the number of foggy days, cloudy days, and days in which the northern lights were visible. He maintained an interest in the state of ice, jotting the presence of cracks and the movements of the ice pack from shore.[28]

The International Polar Year, then, transformed the Arctic environment. Once the northern limit of the fur trade, the Beaufort Sea coast was now the southern edge of the commercial whaling industry. Moreover, both whalers and scientists could operate within this environment on their own, in direct competition with Inuit, by refining knowledge about marine biology, hydrology, and meteorology. When scientists and military officials confirmed Point Barrow as a safe harbor on shore, the expedition solidified this change. The federal government delayed appropriations for a relief station there, but private enterprise moved forward with its own plans. In 1884, the Pacific Steam Whaling Company established facilities near Barrow, Alaska. With the structures of American empire firmly in place, a monument of preindustrial global history—Kittigaaryuit—would soon be leveled.[29]

RECOMBINING HUMAN AND NONHUMAN ECOLOGIES

Over the rest of the 1880s, the commodities that built the whaling industry faced fierce competition from a suite of cheaper alternatives. Hog oils, camphene, and petroleum could be processed and shipped in the south, as opposed to the expensive Arctic whale products. The high price of baleen—$5.38 a pound in 1891, up from 80 cents in 1870—propped up the Arctic whaling industry. These conditions prompted whalers to both intensify and adapt the whaling industry, which extended the recombination of human and nonhuman ecologies on the Beaufort Sea coast.[30]

Whalers first adapted to slimmer profit margins by lengthening the whaling season. Whalers found that by carrying enough provisions to last an entire year, they could access the migration of bowhead whales from the opening of ice in April—a full two or three months before open-water season began on the Beaufort Sea. Staying in the Arctic for the winter required a lot of support. Over the 1870s and 1880s, the industry fashioned a new go-between, tenders, which resupplied active whaling ships during the summer months and took existing cargoes south, preventing any losses that might occur from unforeseen disaster. Like tenders, certain harbors along the Alaskan coast became instrumental in repairing and restocking boats. Port Clarence offered a safer, more sheltered port than others farther northward, like Point Hope (figure 3). As testament to these changes in the whaling system in southern Alaska, whalers stayed longer and more regularly in the Arctic. In the 1890s, whalers made more than seventy winterings on the Beaufort Sea coast.[31]

Wintering-over created new rhythms and exchanges along the Beaufort Sea coast, too. Inuit had developed technologies and traditions around shore whaling—the art of harvesting whale from the melting ice—and whalers were quick to adopt and appropriate these into the commercial industry. As whalers moved northward from San Francisco in June, they hired Inuit families in coastal Alaskan communities that had recently been enrolled as resupply stations, like Cape Prince of Wales (figure 3). They employed Inuit as seamstresses, shelter builders, and caribou hunters, since the long winter required a viable source of protein for the crew that stored provisions could not supply. In return for their year's service on the boats, whalers paid Inuit men and women in trade goods, whaling gear, and whaleboats, providing Native northerners with enhanced mobility. By the 1880s, the Point Barrow village included whalers from all over the world and Inuit from all over Alaska.[32]

Wintering-over also restructured the geography of settlement on the Arctic slope. Over the 1880s and 1890s, permanent villages comprised of whalers and Inuit emerged along the Alaskan and Beaufort Sea coast, stretching from the Bering Strait to Point Barrow. These new towns—including Cape

Lisburne, Point Lay, Point Hope, Wainwright, and Point Barrow—were often sited at existing Inuit encampments (figure 3). In some cases, though, these settlements obsoleted other historic meeting places. Where Inuit fur traders had previously congregated at Sheshalik near the Seward Peninsula in the summertime, they abandoned this site for trade at Barrow by the early 1880s. The site of Nigliq, which had acted as a tributary linking two main channels of trade on the Mackenzie River and Bering Strait, gradually lost its prominence as trade shifted by 1887 to a site closer to Point Barrow called Pigliq (figure 3).[33]

This pattern of searching for whales, establishing trade with Natives, wintering-over, and settling trading centers along the coast led to the US whaling industry's entrance into Canada. In 1889, whalers from Barrow explored the mouth of the Mackenzie River with protection from the US Revenue Cutter Service. Finding whales "as thick as bees," near Herschel Island, they established the island as a hub of trading activity. In 1894, fifteen whaling vessels called at Pauline Cove on Herschel Island, and the town boasted a population of one thousand people (figure 3). By the time the whaling industry came to Canada, markets had worsened, inspiring additional adjustments in operations. By 1890, whale ships took the bulk of their profits from the sale of flour, firearms, alcohol, and other American goods, effectively becoming floating trading posts. When they did, they displaced the largest Inuit settlement in North America, Kittigaaryuit.[34]

The vagaries of economic geography and biology had spelled the end of the Fort Anderson post in 1866, the British empire's attempt to colonize Inuit fur country. They also fueled the continued expansion of the whaling industry along the Arctic littoral in the 1880s and 1890s. Whalers emptied the Beaufort, reducing the population of bowheads from thirty thousand in 1880 to ten thousand in 1910. Whaling boats also brought cheaper and more extensive loads of American-made goods than other existing fur posts offered. A 100-pound sack of flour, for instance, fetched $30 at Fort McPherson, but cost only $2 from a whaler. As whalers and traders from Russia overharvested whales, walrus, and seal in the 1850s and 1860s, Inuit from the Bering Strait became more interested in acquiring Beaufort Sea products, whether by temporarily traveling there to trade or permanently relocating. It is also possible that declining caribou populations along the Bering Sea coast—which were driven by Inuit contracted to procure fresh meat for whalers—further motivated Inuit to head north for the goods they once could find at home. And finally, a series of disease epidemics plagued Inuit as they engaged whalers. One archaeologist estimated a population of ten

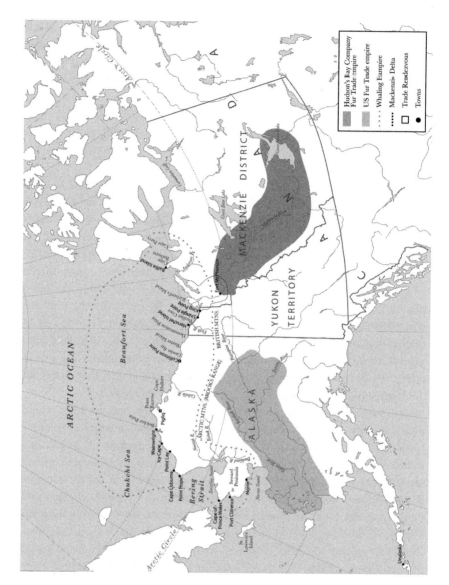

FIGURE 3. The western Arctic in 1898. The arrival of commercial whaling to the Beaufort Sea coast altered the region's settlement geography and major thoroughfares. When combined with the purchase of Alaska by the United States in 1867 and the transfer of northern territories from England to the Dominion of Canada by 1880, the political and imperial boundaries of an earlier colonial era had been reshaped. Map by Morgan Jarocki.

thousand Inuit in Arctic Alaska prior to the arrival of commercial whaling. By 1890, the scholar suggests, that number had been slashed to three thousand. No doubt, sickness encouraged a deeper reliance on the marine trade and hastened the escape to potentially healthier frontiers to the north and east. It also cleared space for whalers and scientists to extend their authority over Arctic waters.[35]

With Herschel Island and Barrow established as anchors in the Beaufort Sea trade, Canada faced a unique challenge to its Arctic dominion. In 1870, three years after the United States purchased Alaska, and the confederation of Canada, the Hudson's Bay Company surrendered large swaths of British North America to Canada. In 1880, Britain passed over its remaining Arctic lands—the islands above the mainland of Canada—out of fear of American intrusion. By 1892, these fears had been realized. In that year, John Firth, the post operator at Fort McPherson, received a party of Inuit with a boat loaded with goods purchased from American whalers, a reversal in the flow of goods and people that had defined the post's life in an earlier empire. Between 1891 and 1907, the value of whales extracted from the Canadian portion of the Beaufort Sea by American whalers totaled $13.5 million. This far surpassed the value of the Canadian fur economy during the same period, which stood at $1.5 million. Herschel Island thus became a crucial meeting ground among the US whaling industry, Inuit, and the Hudson's Bay Company's fur trade in Canada (figure 3). It was a joint that connected and recombined life in the Arctic, whether ecological or imperial.[36]

A STRANGE CALM

As they interacted with coastal communities for longer stretches of time, US whalers and scientists changed their thinking about Inuit life and Arctic land. Ideas of danger that had for so long characterized the circumpolar world began to fade by the 1890s. Such was the final knell of British and Russian empires and a signal of a century of intervention to come.

The research party of the International Polar Year initially reproduced British representations of Inuit as hostile. The commander of the party, Lieutenant Ray, prohibited expedition members from visiting Nuwuk and Utkiavik, two Inuit communities in the vicinity of Point Barrow, because he regarded the Natives as savage. Ray's fears stemmed from what he described as a growing local dependency on the whalers. He noted that Inuit were "disaffected" because their "usual supply of whisky, arms, and ammunition from the whaling fleet" had been unavailable because of a poor whaling year. He also appealed to his superior, General Hazen, for Revenue Cutters to peruse the area and prevent the sale of "contraband goods." These appeals were representations that attest to a tenuous administrative presence in the area. They also document, without intention, the movements of Inuit from western coastal Alaska to the north slope.[37]

Naturalists John Murdoch and Middleton Smith also portrayed Inuit as uniquely powerful. For instance, Smith's notes on the ice conditions of the Beaufort Sea often included remarks about Inuit who were hunting seal or bear. While such activity does not seem to present risk to Smith, the orders that guided his fieldwork laced these observations with concern. Because of the commitment to meteorology and hydrography, and the restrictions from Lieutenant Ray, neither Murdoch nor Smith could venture far from the station. Reading Smith's logbook, one can imagine the experience of being shut in for months on end. Such stillness would have made readily apparent the movement of life around him, a kind of freedom he did not enjoy. Smith showed excitement and trepidation in brief visits from Inuit, who came to the station to trade with him. These visits must have been nerve-racking for Smith and Murdoch because they constituted both a test of Ray's suspicions of violence and the scientists' only chance of procuring natural history collections.[38]

As whalers wintered-over in the north and became inhabitants of the Arctic, their ideas shifted, presenting the region as more friendly as long as one adapted to it. In his autobiography, American whaler-turned-Arctic resident Jim Allen recalled learning how to build houses throughout the year. This involved finding the right kind of snow in winter and collecting driftwood from tributaries in the summer. Allen also described the importance of dogs to the project of traveling, providing readers with instructions on how to harness the animals and what to feed them to keep them strong and obedient. Since dogs fed mostly on seal, Allen and other whalers mastered the seal hunt, including what weapons to use and how to make them, when to hunt seal, and how to hunt in various rain and snow conditions. Residential knowledge, which derived from the experiences of living in the environment for longer than a whaling season, seemed to challenge Allen's former conceptions of the treacherous northern landscape. Of course, his autobiography and that of other whalers who shared their stories in print later in life had ghostwriters. Inuit men and women, who offered Arctic education to American transplants, were often silent and unnamed characters in whaler memories. "When we first went north," Hartson Bodfish remembered, "we provided ourselves with heavy woolen clothing of all kinds."

"But before the end of this first winter," he recalled, "we adopted the native dress altogether, and provided outfits for all the men." In between the lines of Allen and Bodfish's sentiments one can read the labor of Inuit men and women as well as their relationships to Arctic representations in the 1890s. Inuit were being excluded from the north even as their excluders relied on them for survival.[39]

Scientific knowledge also evolved over the 1890s, but not in the same ways as whaler knowledge. Rather than earn a sense of comfort with deeper understanding, scientists developed greater concern, especially over the impact whalers had on northern lands and people. A visit to Herschel Island in 1892 startled naturalist Frank Russell. The Dene guides he traveled with dropped their Gwich'in language when speaking with coastal Inuit, adopting instead a "trade jargon" that included Hawaiian words. Stepping inside an Inuit tent near Pauline Cove, Russell found "coffee, flour, and syrup" from nearby whaling vessels. Before leaving for San Francisco, Russell made a final collecting trip in search of a ptarmigan in its summer plumage. Traveling with Inuit families who were crossing to the mainland to trade with Dene there, Russell noted that his water taxi was a whaleboat owned by Inuit made wealthy from commercial whaling. These changes must have appeared as stark for Russell, who, in preparation for his trip north, had been reading accounts of British Royal Navy explorers. The dangerous and barren Arctic of the mid-1800s was likely no comparison for the strange calm Russell came to know.[40]

GLIMPSES OF THE FUTURE

Frank Russell's arrival in the Arctic constitutes the realization of a region transformed by commercial whaling. A student at the University of Iowa, he became the first scientist to descend the Mackenzie River to its mouth and reach "civilization" by sailing around Alaska. Of course, such a route was only practical because of both the Hudson's Bay Company fur posts and the whaling infrastructure at Herschel Island, Point Barrow, and San Francisco (figure 3). His encounters thus prefigure some of the possibilities—for science and colonialism in this region—that opened in the twentieth century following a profound transition in global networks over the end of the 1800s.[41]

Russell's research interests mark the place of the Arctic within the growth of scientific institutions in North America over the end of the nineteenth century. The development of universities in the last twenty years of the 1800s

provided a pool of eager young scholars, like him, looking to build a reputation in the field of science and a venue in which those careers could play
out. To these ambitious young men, Arctic Alaska and Canada offered proving grounds for one's future tenure as a professional scientist. Of Alaska, US
government scientist Henry Elliot wrote, "It is a paradise for the naturalist, a
happy hunting ground for the ethnologist, a new and boundless field for the
geologist, and the physical phenomena of its climate are something wonder-
ful to contemplate." These articulations would not have been possible in the
mid-1800s, especially given the disarray of natural history collecting at the
time.[42]

As a scientific traveler affiliated with a university in the 1890s, Russell's
research responded to impulses distinct from his predecessors. He received
orders from the Iowa Board of Regents to obtain specimens of larger mammals, especially the muskox. This species had gained notoriety among
sportsmen, university scientists, and leisure-class urbanites as an icon of a
growing social movement, conservation. The destruction of the continent's
natural resources had been made painfully evident in the United States and
Canada through the near extinction of the bison and passenger pigeon, as
well as the exploitation of the interior's timberlands. As tentacles of US and
Canadian empires reached northern shores in the late 1800s, conservationists construed the Arctic as a place in which rational study and management
of nature would redeem the tragedies of environmental history.[43]

To achieve his objectives, Russell created field plans that diverted from the
work of scientists before him. Russell's routes of travel and his interactions
with Native northerners were not confined to the Mackenzie River District or
a remote research station, as MacFarlane and Murdoch had been. Rather, Russell was able to move throughout the Mackenzie River, the Mackenzie Delta,
and the Beaufort Sea coast, allowing him to conceive of this geography as a
connected region. We get a sense of this through his shipments of specimens.
Before heading down the Mackenzie River to Fort McPherson, and then on to
Herschel Island, Russell arranged for a series of boxes of artifacts and specimens to be sent out of the north via Fort Resolution, one of the Hudson's Bay
Company's posts. Once at Herschel, however, he sent similar goods not down-
river, but across the sea, via San Francisco.[44]

Importantly, the whaling industry—and the recombination of human
and nonhuman ecologies it inspired—changed how collectors like Russell
conducted their business. Given MacFarlane's occupation in the fur trade, he
relied solely on Inuit for knowledge and specimens. But the establishment
of regular schedules of transportation and trade on the Beaufort Sea coast

provided Russell with opportunities to perform a more comprehensive study without Native informants. At Herschel Island, Russell found whalers to be a great source of information about an expanse he could not possibly cover on his own. He plied them for testimony on a variety of subjects, such as the conditions of the shoreline, the presence or absence of Native communities along the edge of land, and the habits of ice, geese, fox, and fish. Beyond accounts and observations, Russell also relied on whalers to help him acquire specimens. He outsourced some of his collecting duties to crewmen aboard the *Jeanette*, a whaling vessel that employed both white men and "Cogmollicks," or Inuit from the Mackenzie Delta area. Whalers did not fully displace Inuit as keepers of Arctic knowledge, but they allowed itinerant scientists more access to the northern world and introduced competition within the various exchange relations on the Beaufort Sea coast.[45]

The Iowa ethnologist's movements also reveal an undercurrent operating within the circuit of knowledge and power in the Arctic during the 1890s: Canadian anxieties. By 1895, whalers recognized dwindling bowhead populations around Herschel Island and pushed the edge of the whaling frontier eastward to Baillie Island, even deeper into Canadian territory (figure 3). The presence of American whalers in the Canadian north represented a threat to sovereignty, and missionary reports of debauchery, prostitution, and the free flow of alcohol gave further cause for concern. In 1903, Canadian officials dispatched northern police to Herschel Island to monitor whaling activities and collect customs duties from ships arriving from Barrow, Alaska.[46] Unlike its neighbors to the south, though, Canada was unprepared to dispatch homegrown naturalists to the Arctic in the 1800s. The Dominion boasted fewer universities and museums and did not complete its national museum until 1907. Despite its lack of internal scientific capacity, the Canadian government demonstrated its interests and worries about the Arctic by sending a field-worker to the northwest corner of the country. In 1889, the government contracted a French geographer, Edouard de Sainville, on a five-year study of the Mackenzie Delta and Beaufort Sea. In 1893, de Sainville met Frank Russell, and the two men worked side by side for a year, each accounting for the resources and geography of the Arctic, but for different nations.[47]

In Russell and de Sainville's travels, then, we glimpse both the completion of a profound transformation in Arctic life over the end of the 1800s and shades of the Arctic's future. Developments in research institutions and disciplines continued to make the north a hotspot for science. The rise of conservationist ideologies and international geopolitical tensions compelled scientists to study and steward the Arctic. Opportunities for comprehensive fieldwork across the region emerged only because of economic expansion there. And resource exploitation displaced Arctic residents and Arctic nature, even as it was conditioned by a host of interactions among Inuit, transportation technologies, biological invaders, and northern ecosystems. All of these trajectories intersected over the next hundred years, making and remaking the Arctic, and thrusting the place out of the shadows of an earlier imperial age.

CHAPTER TWO

Threatened: The Ambitions and Anxieties of Expeditions

In the summer of 1913, an unlikely crew sat aboard the *Mary Sachs*, stuck in an ice-choked harbor in Nome, Alaska. The 30-ton, 60-foot whaling schooner boasted none of the familiar deckhands and seamen from San Francisco ports. Rather, the vessel housed a crowd of Alaskan Inuit and a handful of professional scientists from Canada, Denmark, the United States, and Australia. These were the members of the Southern Party of the Canadian Arctic Expedition, the largest and most expensive government-sponsored scientific expedition ever to study northern North America. Two geologists, a geographer, a marine biologist, a zoologist, and an anthropologist—all with PhDs—composed the Southern Party. A second group, the Northern Party, featured an anthropologist, a geologist, a meteorologist, an oceanographer, and one of the world's decorated explorers, Vilhjalmur Stefansson.[1]

The *Mary Sachs* was a harbinger of Canadian colonial relations in an Arctic long dominated by Russia, England, the United States, and Inuit communities. Of the nine ships waiting for ice to shove off the coast in 1913, three pursued new lands and scientific knowledge for the Dominion, not the baleen or blubber of bowheads. The expedition lasted five years, ran a tab of more than a half million dollars, and cost the lives of nearly a dozen scientists and crew on the lost expedition vessel *Karluk*. The *Mary Sachs* became one of the first vessels to fly the Canadian flag while crossing the international border at Demarcation Point, Alaska, signaling Canada's sovereignty as leading nations raced for the Poles.[2]

Canada's foray into Arctic affairs in the 1910s constitutes a landmark in a century of science, intervention, and transformation. Not simply because it heralded a major salvo in an intense global competition, but because the

Canadian government deployed professional scientists in this campaign. In the years before the Great War, imperial development hinged on territorial expansion—whether through discovery, conquest, or purchase. Expeditions became, for the Dominion, another means of ascension. Between 1899 and 1918, federal agencies, natural history museums, and wealthy philanthropists dispatched four expeditions to the Beaufort Sea coast. While scientists had traveled northward with such support before, expeditions in the 1900s moved differently than earlier scientific activities. They were both extensive and intensive. That is to say, they covered large swaths of land and performed deep analyses of certain areas. Such coordinated movement draws attention to how Canadian anxieties and ambitions—an undercurrent in the late 1800s—surged northward by the 1910s.[3]

Because of their cost and distance from centers of government, expeditions open questions about the forces weaving science and colonialism together. What did scientific communities hope to learn about the far north? Why did Canada invest so heavily—whether financially or politically—in the Arctic in the first two decades of the 1900s? While scientific sponsors advocated for northern exploration and resource exploitation, they also worried about industry's long-term consequences on Native cultures and the land. Alongside scientists in search of discoveries, then, other field-workers pressed north as archivists—and both embodied the politics of colonialism in the early 1900s. Geographers, geologists, and explorers searched for a potential "Polar Continent" and valuable mineral deposits in the region. Anthropologists hoped to document traces of so-called uncontaminated indigenous peoples. Expeditions also make clear the ways Arctic life—both human and more than human—mediated colonial desires. Expeditionary science would have been impossible if not for the landscape changes involved in establishing coastal outposts through the commercial whaling industry, or the widely distributed Inuit and resident whalers who guided scientists through northern landscapes. This is especially true at the turn of the twentieth century when the Arctic did not yet know the wireless radio or the airplane.

In these ways, expeditions conjoined social concerns, political dreams, academic interests, industrial practices, and ecological realities particular to the early twentieth century. It is no coincidence that this style of survey reached its apex in these years as American, Canadian, and European powers sought to capture the world's "unclaimed" lands, often in competition with one another. Hints of this history can be found on the *Mary Sachs*, the former whaling schooner outfitted for state-sponsored scientific investigation in 1913. From the exterior, the boat manifested as another whaling vessel. In actuality, it was filled with unique imperial desires, a distinct

FIGURE 4. The Canadian Arctic Expedition schooner *Mary Sachs*, near Herschel Island, Yukon Territory, 1914. Photo taken by Kenneth Chipman. Credit: Canadian Museum of History, 43249.

epistemological community, and the potential for another environmental transformation in the Arctic (figure 4).

THE MYSTERIOUSLY MODERN CONDITIONS OF WESTERN ARCTIC EXPEDITIONS

Historians have emphasized the search for the North Pole as part of a global contest for the earth's remaining resources at the turn of the twentieth century.

Far less known, but equally as important, are a set of four contemporary expeditions to the western Arctic. These Western Arctic Expeditions were after a supposed Polar Continent, over which no nation had yet declared ownership, and a "tribe" of Inuit who had apparently never before seen white men—yet embodied certain traits of "white civilization." Mysterious lands and peoples galvanized global scientific and colonial interests, which eventually helped rationalize Canadian investment in the Canadian Arctic Expedition in 1913.[4]

In the early 1900s, American geologists and oceanographers posited that undiscovered land might exist offshore of northern Alaska. Studying the observations and travelogues of polar explorers from the mid-1800s, these scholars pointed to the directions of currents and tides in the Beaufort and Chukchi Seas, which suggested a landmass just above today's Alaska-Yukon boundary. Published in *American Geologist* and *National Geographic Magazine*, the news stirred geographers and explorers. The alleged Polar Continent turned out to be a mirage—a trick of the eye that mistook bumpy sea ice for firm ground. It was one of many phantom islands that appeared in the accounts of explorers as Norway, Germany, England, the United States, Russia, and Denmark combed the circumpolar region. Nevertheless, the illusory continent created real consequences. As exploration eliminated "unexplored" spaces from the world map, the value of those surviving increased. British geographer Clements Markham proclaimed in the *Geographical Journal* that the Beaufort Sea was "the least-known part of the Arctic Regions, and the one which contains the most interesting geographical problems." Markham considered this corner of the Arctic as the last stone to be unturned in a grand edifice of circumpolar knowledge, whose construction had begun centuries before.[5]

Markham's promotional efforts can be tied to two of the four Arctic expeditions we follow in this chapter. As a leading figure in the Royal Geographical Society, Markham sponsored explorer Alfred Harrison to "penetrate as far as possible" into the Beaufort Sea in hopes of discovering "whether there was land hitherto unknown in the Arctic Ocean." As Harrison headed north, another expedition had already arrived there. In 1906, Danish explorer Ejnar Mikkelsen and American geologist Ernest de Koven Leffingwell formed the Anglo-American Expedition. While this team shared Harrison's intent to find the "undiscovered continent," they did not share the same motivations. The Anglo-American Expedition was, contrary to its name, not a British scheme, or even a joint venture of British and American sponsors. Rather, its bankroll came from a mishmash of philanthropists, royal elite, publishers, scholars, governmental officials, corporations, and relatives of the expedition's leaders.

The diversity of funding sources indicates the expense of an Arctic expedition. But it also suggests the purchase of a Polar Continent to industrial society. It appealed to scientists, the reading public, leaders of nations, and the wealthy elite in the early 1900s because it represented the human conquest of the Earth.[6]

Neither Harrison nor the Anglo-American Expedition found the supposed offshore land. Many scientists grew to doubt the island's existence, while others maintained that the limited measurements taken by scientists on the Anglo-American Expedition could not disprove the Polar Continent theory. In a peculiar way, failures of science beget renewed commitments to science. Harrison, Mikkelsen, and Leffingwell thus helped stimulate two more expeditions, the Stefansson-Anderson Expedition of 1908–12 and the Canadian Arctic Expedition of 1913–18.[7]

Importantly, the Polar Continent formed only one component of designs on the Arctic. In addition to discovering new lands and riches, scientists sought to document forms of northern life that were thought to be obliterated by expanding industrial economies. As a member of the Anglo-American Expedition, anthropologist Vilhjalmur Stefansson heard rumors of an "uncontaminated" tribe of Inuit in the eastern stretch of the Beaufort Sea, who seemed to have European features—like red hair and blue eyes—and a language that included words from Old Norse. Stefansson wondered if this native race explained the fate of the lost Norse colonists of Greenland, a mystery that had perplexed archaeological communities during the nineteenth century and became fodder for adventure novels in Europe and North America over the turn of the twentieth century. Stefansson channeled his excitement about these Inuit toward the American Museum of Natural History in New York, with which he collaborated to establish the Stefansson-Anderson Expedition of 1908–12. The anthropologist-explorer teamed with fellow University of Iowa alum and zoologist Rudolph Anderson, who saw the expedition as an opportunity to generate full collections of Arctic species that were poorly represented in existing North American museum holdings. Stefansson and Anderson spent four years in Alaska and northwestern Canada, recording the biological and cultural life of the region and shipping tons of specimens and artifacts to the American Museum in New York. Anderson's zoological studies—though important finds for science—were subsumed by Stefansson's discovery of the "Blond Eskimos." The *Seattle Times* and *New York Times* promoted Stefansson's contact with these Inuit as an incredible ethnological contribution, one that proved the existence of a Stone Age people in the civilized world.[8]

Set in the context of their shared historical moment, the fanfare of the "Blond Eskimos" and the Polar Continent smacks in similar ways. Both were misleading scientific representations that nonetheless galvanized intervention and environmental transformation in the Arctic. Even more, they reflect an early twentieth-century preoccupation with the decay of nature and society. Caught in a period of rampant urbanization and immigration, North American city dwellers found Darwinian evolution a salve for social disorder. This theory understood the development of all life as fueled by a competition for scarce resources. As time rolled on, so did natural selection: those species unable to secure their own livelihoods would die, leaving only the fittest to survive. Metropolitan observers, then, imbued phantom islands and lost peoples with special meaning. New continents could secure a nation's land base, the material basis for population growth, wealth, and power. And "strange looking peoples" confirmed important ideological foundations of Western civilizations. As anthropologist John L. Steckley has written, the implications of the "Blond Eskimo" fortified American and Canadian conceptions of their own evolution as a story of superior technology and tenacious martial spirit. The presence of Norse descendants in the western Arctic suggested that the "survival capacity developed through hundreds of years of Inuit cultural innovation could readily be equaled by Westerners in a short time." Threatened peoples and places in the Arctic were thus products of a mysteriously modern condition.[9]

When convened in 1913, the Canadian Arctic Expedition of 1913–18—the last expedition in this group of four—took as its assignment the final resolution of the Polar Continent and "Blond Eskimo" controversies. Reflecting the importance of the far north to science and global power, and Canada's imperial dreams, this Arctic expedition was the largest and most organized of any in its day. It featured two groups of researchers: a Northern Party, led by Stefansson and in search of the Polar Continent and other undiscovered lands in the Beaufort Sea, and a Southern Party, commanded by Rudolph Anderson and charged foremost with investigating reports of copper deposits in the Coronation Gulf. The government also tasked the Southern Party with mapping the Mackenzie Delta, bringing home serial collections of all terrestrial and marine life, and documenting the material culture of all Inuit communities of the region, especially the "Blond Eskimo." In its size, composition, and timing, the Canadian Arctic Expedition crystallizes the ambitions and anxieties of colonialism on the eve of the Great War.[10]

CARRYING OUT EXPEDITIONS,
EMBODYING EMPIRE

Harrison's Expedition, the Anglo-American Expedition, the Stefansson-Anderson Expedition of 1908-12, and the Canadian Arctic Expedition of 1913-18 constitute an unbroken stream of scientific activity in the western corner of the North American Arctic. These Western Arctic Expeditions earned unprecedented support from the broader public and scientific communities to ensure industrial society's future, through claiming for their own purposes the natural resources and cultural value of foreign territories. Importantly, notions of the Polar Continent and the Blond Eskimo also provided a pretext for Canada to assert itself against other world superpowers.

Canadian officials viewed the Canadian Arctic Expedition as a chance to end their bullying by England and the United States. That the Dominion struggled with these rival nations to assert its claims to the north had been made painfully clear during the search for gold over the turn of the century. In 1898, with ports in Seattle and San Francisco buzzing over reports of gold strikes in the Yukon Territory, more than twenty steamers left the American West, carrying with them nearly twenty thousand stampeders. Canadian officials worried about so many Americans on their own soil and the profits they took with them out of the country. Importantly, the Klondike gold rush catalyzed the establishment of the international boundary line between Alaska and the Yukon—the 141st meridian—in 1903. Since England handled Canada's foreign affairs at the time, London led negotiations with President Roosevelt. The two nations each appointed three members to a tribunal, which, through defining the boundary, granted the United States much of the Alaskan panhandle and the key seaport of Skagway. Historians conclude that this decision was a gift from England to the United States to maintain peaceful relations. This agreement set the United States and Canada on divergent paths within the north. While Americans focused on strengthening the economic corridor between the mineral lands of the Alaskan interior and the timberlands of the southeast, Canadians felt their national development had been sacrificed for the benefit of England. They looked to the Arctic to prove themselves on the world stage.[11]

These international tensions are evident in the correspondence between Canadian officials and US scientific leaders over the Canadian Arctic Expedition. Vilhjalmur Stefansson initially agreed to run the expedition through the National Geographic Society and American Museum of Natural History as the "Second Stefansson-Anderson Expedition." Provoked by the border conflict with the United States, though, the Canadian government stepped in. They understood that an expedition out for new lands and "lost" tribes could not only suggest their political capacity but also materialize it. Extensive copper deposits were rumored to exist in the land of the "Blond Eskimos," or as they were also known, the "Copper Inuit." Canadian officials informed Stefansson's existing sponsors of their intent to take over the initiative, renamed it the Canadian Arctic Expedition, and outfitted it through their Geological Survey and Department of Naval Service. They also planned for the collected specimens and artifacts to be housed in a new national museum, the Victoria Memorial Museum.[12]

As this exchange shows, the Canadian government sought to define its sovereignty through scientific practice. Traditionally, claims to northern lands were made through the language of treaties or material possessions—like police posts, customhouses, post offices, or hospitals. Canada maintained few of these in the western Arctic, which compounded the uncertainties of its foreign affairs situation there.[13] In 1903, Canada had sent a Dominion Government Expedition to solidify claims in the eastern Arctic—specifically around the northern shores of Hudson Bay, where American whalers also operated. Federal officials similarly authorized the Canadian Arctic Expedition. The *Mary Sachs*, the boat on which we journeyed to open this chapter, was part of the expedition's fleet of five schooners known as the "The Expedition Navy." Government sponsors tasked expeditionary scientist Kenneth G. Chipman with taking duties from American whalers and enforcing customs laws if he saw infractions along the Beaufort Sea coast.[14] Yet the Canadian Arctic Expedition went further to use scientific information and objects as political tools. The expedition became the first coordinated scientific activity to any part of the Canadian Arctic wherein all collections remained within the country. Specimens were deemed "government property," and the observations scientists gathered became items of national security, too. At the same time, the politics of possession in the 1910s encouraged Canadian officials to dismiss Inuit—and their long history of settlement—as a possible avenue toward sovereignty.[15]

As the decision of the international boundary tribunal encouraged Canadians to become more aggressive in the Arctic, it also made United States dial back there. With the panhandle of Alaska now squarely within the nation, the US Geological Survey focused its efforts on the expansive mineral districts in the southern parts of the territory. As long as that region remained productive, the survey had little incentive to continue to fund exploratory work elsewhere. Meanwhile, the American Museum of Natural History and National

Geographic Society shifted their approach in the Arctic to Greenland, where Roosevelt and other bureaucrats hoped to extend American influence.[16]

These relationships between the United States and Canada clarify the activities of Ernest de Koven Leffingwell, who remained in the Canning River region of Arctic Alaska for seven years after the Anglo-American Expedition of 1906–7. Geologists with the survey lamented the fact that the widely distributed knowledge and experiences of gold prospectors was not immediately available to US officials. In 1904, the survey published a map of Arctic Alaska indicating the eastern and western portions as "unexplored." The eastern section of this map aligned perfectly with the area that Ernest Leffingwell studied between 1907 and 1914. Accordingly, the US Geological Survey showed great interest in his geographical and geological research, eventually publishing his results. Leffingwell had thus provided the nation with rigorous scientific knowledge of a little-known part of Alaska at practically no cost to the federal government. When expeditions moved, they reflected the course of nations—and effected them, too.[17]

THE PROFESSIONALIZATION OF ARCTIC COLONIALISM

The rise of the Arctic as such a crucial juncture for science and political economy—let alone Western civilization—opened questions about who was best authorized to make new knowledge. At the turn of the twentieth century, for example, Canadian geologists saw the efflorescence of northern fieldwork following the Klondike gold rush as insufficient, even though it represented the most recent period of data collection. The director of the Geological Survey of Canada, George M. Dawson, argued that the northern expansion of resource economies had been followed by scientists with few capable instruments and little professional training. For Dawson and other government scientists, a thorough understanding of the Arctic territory and its resources could only be achieved through scientific agents with the highest levels of university education and the best scientific equipment. It is not surprising that just a few years after Dawson's indictment of amateur collectors, the Geological Survey of Canada, like its counterpart in the United States, mandated all of its staff scientists hold PhDs.[18]

Frank Russell, the Iowa ethnologist who noticed American empire spilling over into Canadian territory in the 1890s, became one of the first victims of new standards for American and Canadian science in the early twentieth century. Consider reviews of his book, *Explorations in the Far North*, in two different journals, the *American Naturalist* and *American Anthropologist*. The first review lauded his work as an important contribution to science. The second, by respected American physical anthropologist Arles Hrdlicka, condemned the research as nontechnical and suited more for popular audiences interested in adventure. The mixed reviews indicate the ways scholars created new protocols in the early 1900s for research and publication. These protocols bonded scientists with particular training into specialized communities, which were identified through formal societies and academic disciplines. These same guidelines excluded individuals who did not have requisite training or social status. As field-workers and scientists fashioned themselves as professionals, they relegated others as "amateurs." Russell's reception thus makes visible the emergence of a professionalizing scientific society in and outside the Arctic.[19]

To get a better sense of how different standards for research practice shaped Arctic colonialism in the early 1900s, we can turn to the Anglo-American Expedition. Its leaders, Ernest de Koven Leffingwell and Ejnar Mikkelsen, represented two different traditions of scientific travel. Leffingwell was a trained geologist, Mikkelsen a self-proclaimed explorer. To attract readers, explorers often regaled sensationalistic accounts of their journeys, which cashed in on the age-old appeal of the Arctic as a death trap. Mikkelsen wrote *Conquering the Ice* (1909), for example, even though his expedition was unsuccessful in finding new land. Leffingwell, who studied geology at the University of Chicago, denied interviews to journalists who sought typical Arctic stories. Leffingwell described the role of science in the Arctic in an editorial in the university's alumni magazine. "I had not been engaged in gathering material for popular presentation," he explained, "nor had I many tales of hardships and sufferings to relate." In fact, Leffingwell continued, he had purposely avoided such tales, because he considered them "the result of carelessness or incompetence." Leffingwell's comments echo the analysis of historians of exploration, who have maintained that the accounts of explorers were a foil for the values of rationality and objectivity—which came to define scientific practice in the early 1900s (figure 5).[20]

In order to be taken seriously by colleagues and governmental sponsors, scientists had to manage more than just how they spoke about their travels. They also had to follow emerging expectations for collecting data. As a testament to his commitment to scientific rigor—and not pure adventure— Leffingwell completed his survey of the Alaskan coast in the methods outlined by the US Coast and Geodetic Survey and the US Geological Survey. This demanded discipline on Leffingwell's part. When harsh weather, months of darkness, and failed deliveries jeopardized his fieldwork, he did

FIGURE 5. Ernest de Koven Leffingwell, standing before his astronomical observatory on Flaxman Island in 1906, during the Anglo-American Expedition. The different commitments of scientists and explorers, and the entanglements of those commitments, are symbolized with Leffingwell's use of discarded boxes from Horlick's Malted Milk to protect his research station. Horlick's was a sponsor of the expedition. Credit: US Geological Survey.

not waver. Instead, he returned to his field station on Flaxman Island for five summers and three winters, finishing his research over a seven-year period. During that time, he realized his equipment malfunctioned in the bitter cold and thus had custom instruments built for him to ensure accurate measurements. Like other young scientists, Leffingwell was incentivized to go to these lengths because he sat on the bubble of professionalizing academic communities. He was not a governmental employee, and he had abandoned his PhD studies to take part in the Anglo-American Expedition. Leffingwell was thus "much relieved" when Alfred Brooks, the director of the US Geological Survey, recognized the character of his Alaskan studies and committed to publishing it. Unlike Frank Russell, Leffingwell earned the approval and status of a professional scientist.[21]

Over the early 1900s, then, governments interested in exacting political and economic interests on distant shores often turned to a select bunch of researchers, not simply those already stationed in a foreign land or those intrigued by a sense of danger. Again, the Canadian Arctic Expedition, known widely as boasting "the largest scientific staff ever taken on an Arctic expedition," represents the culmination of this professionalization trend. One of its leaders, Rudolph Anderson, noted that an expedition differed from other forms of scientific travel because it required intelligent, specially trained, young, and ambitious men. They were recruited not by enormous salaries or book contracts, but by the "desire of accomplishing some individual work which would do credit to himself and aid him in his future work." Among the more than two-dozen specialists hired to take part in the Canadian Arctic Expedition, the majority were younger than thirty years of age and had just finished their doctoral degrees in the sciences—ready to launch their careers. By following a "regular scheme" for fieldwork, the Canadian government could rest assured that collected knowledge was of a "high order," enhancing both the control of nature and their chances of success in a Social Darwinist struggle for imperial existence.[22]

THE ECONOMIES OF EXPEDITIONS

Regardless of their training or age, though, scientists could not move in the Arctic on their own. The maintenance of economic infrastructure—outposts, thoroughfares, and ports—facilitated the activities of expeditionary scientists in the region in the early 1900s, even as the whaling industry bottomed out. In an odd way, then, the Canadian Arctic Expedition could only attain its goals of political and economic expansion by relying on economies set up by American whalers and Alaskan Inuit.

In 1907, the arrival of plastic substitutes for baleen undercut the whaling industry. While many whalers discontinued their yearly trips to the Beaufort Sea, the whaling vessels themselves did not disappear. Some American whalers converted their operations to the procurement and sale of Arctic fur. The most successful of these transformed their whaling stations into fur posts and their whaling vessels into floating stores. Because fur-bearing animals and whales have different habitats and life cycles, the bulk of harvesting activity shifted from the open waters to the coastline and from the summer to the winter months. Despite these shifts, the open-water season remained the critical period of exchange. The Beaufort Sea coast also remained the home of trading relations, especially at places like Point Barrow and Demarcation Point—the border between the United States and Canada—as well as Canadian locations like Herschel Island and Baillie Island.[23]

The return of a fur economy in the Arctic steadied the wake of the whaling industry in other ways. Biological invaders accompanying whalers took a toll on Inuit. In 1850, an estimated 2,500 Inuvialuit inhabited the Beaufort Delta region. By 1910, after epidemics of measles and Spanish influenza, only 150 survived. Remarkably, fur trading filled these voids. The eastward push of the fur frontier in the first two decades of the twentieth century contributed to a migration of Alaskan Inuit into the Mackenzie Delta. These Inuit fanned out over the stretch of land between Point Barrow and Baillie Island, repopulating and stabilizing communities there. Residents relied on floating posts for shipments of goods, equipment, and ammunition in the summertime. In turn, these settlements—spaced no more than 15 miles apart from one another—formed a foundation for extensive and intensive scientific research.[24]

The itineraries of Western Arctic Expeditions make visible the importance of trading vessels and Arctic communities in the project of science. Many expeditions began by hitching rides with whaler-traders leaving San Francisco bound for Barrow, Alaska. Once in the Arctic, researchers learned valuable information from captains about how to make the most use of the open-water season, the best time for safe, extensive travel in the north. Field-workers also used the seasonal movements of whalers and traders to ensure communication with loved ones and government superiors. News from the south could redirect activities in the months ahead, as it did for Canadian Arctic Expedition members learning of the Great War or scientists attempting to stay up to date with recent publications in their disciplines. As such, expedition members were acutely aware of the difficulties in sending and receiving mail, often devoting large sections of their official reports to the schedule, cost, and logistics of shipping letters via the Mackenzie River or around the coast of Alaska.[25]

Scientists carefully manipulated their interactions with whalers and traders toward their own professional benefit. In 1910, anthropologist-explorer

Stefansson reached the country of the "Blond Eskimo," supposedly beyond the economic frontier. For the next two years, he stayed within the Coronation Gulf region, zigzagging across islands and the mainland, and visiting a total of thirteen Inuit groups. His field diaries and correspondence indicate that he could operate in this district only because he collected valuable furs, which he traded with locals for food, specimens, and other goods. In the fall of 1910, whaler-turned-trader Joseph Bernard sailed from Herschel Island to Langton Bay, where he held a rendezvous with the scientist. In need of restocking his own supplies, Stefansson exchanged fox skins for a rifle, ammunition, flour, and cloth for a tent. Given that Bernard agreed to ship Stefansson's collections to Nome the following summer, the anthropologist likely promised in return a continued supply of fur. Clearly, Stefansson embraced the trading ships as they fit the project of documenting a threatened Arctic, even as he fingered these economic agents as endangering the Arctic in the first place.[26]

In 1913, when the Southern Party of the Canadian Arctic Expedition arrived in the region, the group made its headquarters at Langton Bay. They promptly renamed this location Bernard Harbour, after the captain Stefansson had met there.[27] Among the team was Diamond Jenness, an anthropologist whose duty included a three-year study of the "Blond Eskimo." Unlike Stefansson, Jenness spent his first year and a half stationed at Bernard Harbour, inviting Inuit in for investigation as these Natives traded with nearby whalers-turned-traders. He also executed a six-month journey with a family throughout Victoria Island, during which he relied on a local translator, himself the son of the whaler who originally told Stefansson about the Coronation Gulf people.[28] Eventually, Jenness relied on these studies—which treated a greater geographic area and examined more individual Inuit—to counter Stefansson's claims about the mixing of Norse and Inuit blood. Jenness found that so-called Blond Eskimos shared more similarities than differences with Inuit across the Alaskan, Canadian, and Greenlandic Arctics. He also noted no instances of red beards (only dark brown ones), and explained the rare blue-eye phenomenon as a product of age and repeated snow blindness.[29] These partnerships, interpretations, and movements (and lack of movements) would have been impossible, let alone imaginable, in the decades before Western Arctic Expeditions.

NEGOTIATING THE ARCTIC

As hinted in the itineraries of Stefansson and Jenness, expeditionary science was not as simple as performing research within the confines of an outpost or from the hinterland of one's tent. Disciplinary norms prescribed certain movements and help explain various activities on Western Arctic Expeditions. Zoologists hunted game across multiple watersheds. Anthropologists traveled only where their Inuit subjects went. Geographers attempted to map poorly known regions by referencing known landmarks. And geologists concentrated on areas rumored to contain mineral deposits. Some of these areas and items were easier to access than others, but all relied on leaving the coast and entering the interior of the Arctic landscape—a banality of Inuit life that became an extraordinary feat for expeditionary scientists. Those who understood the inland Arctic thus acted as intermediaries. In some cases, these go-betweens were Inuit, who regained authority lost over the end of the 1800s with the arrival of commercial whalers to the Beaufort Sea coast. But by 1913, there were other northerners contending with Inuit for residential expertise.[30]

While Rudolph Anderson's seven years in the western Arctic involved a diverse set of experiences as commander, mammalogist, and ornithologist, his habits as a collector can be captured with a photograph (figure 6). Anderson sat alongside his tent at Bernard Harbour in June 1915, his chin tucked to his chest. In his lap, an animal became subject to the demands of science and colonialism; it was turning into a research object and the property of Canada. Next to him, a young boy—his assistant Patsy Klengenberg—adopted a similar posture. Klengenberg, the son of a Danish whaler, spoke fluent English and Inuit languages. He had traveled with Anderson for months, learning how to prepare bird and mammal skins.[31] Anderson's absorption with his scientific work signals an important victory for the zoologist: he was able to look at animals as something other than food. His first winter in the Arctic had brought him illness and near starvation, which had highlighted the conflicting values of animal parts. But soon, Anderson gained assistance that allowed him to see "living off the country" as consistent with, if not beneficial for, his collecting. While in the Colville River area in 1909 and again along the Coppermine River in 1911, Anderson hired Inuit families to take him to areas no previous zoologist had entered. Travel down rivers and into the interior brought him to migrating caribou and into the ponds housing muskrat, marten, mink, and beaver. Anderson mastered the technique of preserving animal materials, which involved proper skinning, treating with chemical baths, stashing in caches atop driftwood frames (to prevent predation), and thwarting the efforts of pesky mice.[32]

Anderson's diary entries confirm his contracts with Inuit were key to the success of the Canadian Arctic Expedition. Anderson hired Ambrose Agnavigak, an Inuk from Herschel Island, in the summer of 1915. For his efforts

FIGURE 6. Patsy Klengenberg (*left*) and Rudolph Anderson cleaning specimens at Bernard Harbour, Northwest Territories (Nunavut), 1915, during the Canadian Arctic Expedition. Photo by George Wilkins. Credit: Canadian Museum of History, 50932.

in hunting, trapping, as well as in building and repairing a 14-foot dog sled, Anderson paid Agnavigak $88 cash. While difficult to know the exact value of that salary, the Inuit man would have been able to convert this payment into a slew of goods at Herschel Island, including a new high-powered gun and a winter's worth of flour, tobacco, and tea. Anderson repeated these relationships with several other men on the Canadian Arctic Expedition, including Ikey Bolt, Adam Ovayuak, Manilenna, and Palaiyak, who were referred to as the "Southern Party Boys." In ways critical for expeditionary work, these guides had familiarity with Native languages, English, subsistence activities, and interior landscapes. These Inuit held the power of being able to transport scientists to places labeled "unknown" on their maps and yet communicate clearly where they were. The size and scope of Anderson's collections reflect debts to his Inuit assistants. Anderson returned to the National Museum in 1916 with over four hundred specimens of mammals, representing over twenty-two species, and six hundred specimens of birds, representing over seventy species. These materials soon became the basis for a series of reports, museum displays, and policies concerning Arctic nature.[33]

Time and again, expedition members convinced Inuit to preserve animals or dispose of personal items by paying them cash and supplying them with foreign goods. Arctic field-workers had few of the resources available to their contemporaries in survey parties elsewhere on the planet, who tapped informal networks of settlers to find and collect specimens. There were, however, resident whalers who had begun to amass their own assortment of curios. In these ways, the social contract linking scientists and local residents negotiated not only the demands on scientists in the Arctic but also the region's cultural and environmental realities. By the 1910s, both Inuit and whaler-traders acted as knowledge-keepers in the inland Arctic, and Canada's desire to defend and develop that place animated the authority of these northerners.[34]

Like Anderson, anthropologist Diamond Jenness used trade—and thus, Inuit labor and knowledge—to document Inuit culture. Jenness walked a fine line between trading for science and trading for profit. His official contract with the Canadian Arctic Expedition forbade him from "trading or other commercial transactions" unless these were completed for "the expedition itself, or the Geological Survey of Canada." But it was nearly impossible for Jenness to avoid trading activities, even if he found other means, like the camera or the phonograph, to record Inuit life. Jenness was the only member of the Southern Party who spent time learning Inuit languages, as well as the trade jargon that became the primary means of communication among whalers, prospectors, traders, and Inuit in the region during the early twentieth century. Jenness constantly toggled between his roles as anthropologist and negotiator, bargaining with Inuit to secure the services and goods the expedition required, such as seamstresses, navigators, boat mechanics, clothes, and food for themselves and their dogs.[35]

Trading was also a means of gathering archaeological artifacts, those items displaced by the very people and economies that made expeditions possible. Ironically, the regional whaling industry had created an archive of the life it had interrupted in the 1870s. Jenness realized that whalers could be rich sources of archaeological curios, since they had been in the north longer than itinerant scientists. While iced-in at Point Barrow in February 1914, Jenness sketched a whaler's collection of Inuit goods, so as to create a database for comparison with items he anticipated gathering farther east. Paying for archaeological items was far easier than digging for them, not simply because of the savings in energy and time. Snow covered the ground for up to nine months of the year, and, for the remainder, the soil was often frozen just inches below the surface. Accordingly, Jenness timed his digs in the summer months and located them at abandoned shelters and grave sites. His specimens became central to the Canadian scientific community's authority over Inuit culture, especially its historical movements and its encounters with Western civilization. Fortunately for Jenness, the emergence of the fur economy—and the shift in harvest seasons from summer to winter—meant that many Inuit were available for archaeological work in the only part of the Arctic year when it was possible.[36]

WHAT THEY TOOK AND WHAT THEY LEFT BEHIND

The picture of Western Arctic Expeditions as conjunctions of processes working at multiple scales resolves when we locate these scientific operations in time. While the planet—and the Arctic—had witnessed scientific travel before, the early twentieth-century version was markedly different. Gone were the days of Roderick MacFarlane, individual explorers or gentlemanly collectors who ventured to a dangerous north in hopes of growing the British fur trade. A more recent era also could now firmly be placed in the past—when naturalists went north with the support of the US government to secure the future of commercial whaling in the Arctic. In retrospect, a lack of suitable infrastructure had kept both kinds of previous encounters as either extensive *or* intensive. Governmental scientists articulated these distinctions at the turn of the twentieth century as symptoms of amateur work, thereby redefining Arctic expertise. By the early 1900s, fur trading reemerged along the Beaufort Sea coast, appropriating and extending rendezvous points, exchange relations, and international tensions created by whalers and Inuit at the end of the 1800s. Because of this history and in order to transcend it, Canada for the first time sent highly trained scientists with PhDs to study and defend the region. Meanwhile, international controversies over a Polar Continent and "Blond Eskimo" swirled, inspiring young researchers to sign on for northern fieldwork, and providing a sheen of enlightened concern for colonial intervention.

Both on their way north and on their way home, expeditions took with them the ambitions and anxieties of imperial nations. Following the vessels of Western Arctic Expeditions, like the *Mary Sachs,* launched this study of expeditions. A similar journey foreshadows what legacies they left behind. When the Canadian Arctic Expedition returned from its five-year reconnaissance in the region, many political and intellectual elites agreed that the task of scouring the north (and the Earth) was complete.[37] Not only had the Northern Party discovered a clump of islands and disproved the Polar Continent theory, but also the Great War erupted in the middle of fieldwork. For at least a short period, scientists retired the expedition as a practice of knowledge production in the western Arctic. Instead, they turned to the load of recently collected Arctic materials. By 1918, members of the Canadian Arctic Expedition delivered to the Victoria Memorial Museum an overwhelming number of scientific objects: one thousand bird and mammal specimens, more than two and a half thousand articles of clothing, hunting and fishing equipment, four thousand photographs, nine thousand feet of film, and ninety-two wax-cylinder recordings of songs and performances. Logbooks and field journals also contained piles of untreated data, like daily observations of weather conditions and a list of tidal observations at seven different locations for three consecutive years. Especially in a postwar world that rejected conquest as a means of imperial expansion, expeditions give new meaning to the ways southern bureaucracies acquired the Arctic.[38]

Like whalers who earned their profits only by selling baleen at market in San Francisco, field-workers processed these raw materials into items of greater value. By the late 1920s, the Canadian and US governments published over sixty reports from the Western Arctic Expeditions, covering the biological, geological, and cultural life of the northwestern corner of the continent. The scientific community hailed these conclusions as groundbreaking. Aldolphus Greeley, the renowned American polar explorer, called the Canadian Arctic Expedition volumes the "most valuable scientific contributions relative to Polar Canada ever published." Alfred Hulse Brooks, the director of the US Geological Survey, referred to Ernest de Koven Leffingwell's 1919 Canning River report as filling in what was a "complete hiatus in the scientific knowledge of Alaska." Federal governments wasted little time in capitalizing upon this knowledge, withdrawing minerals from private staking in northwestern Alaska and in the Coronation Gulf of Canada.[39]

In many cases, scientists working on Western Arctic Expedition materials translated their fieldwork not only into publications but also into careers in civil service. After landing a post in the Geological Survey of Canada, Rudolph Anderson was promoted to chief of biology, a title he held until the 1940s. Anderson was a leading advocate for the conservation of big game species in the north like caribou and muskox through hunting regulations and the creation of parks and sanctuaries.[40] The Geological Survey of Canada hired Diamond Jenness as an anthropologist, and he eventually became the leading anthropologist in the country. Over his tenure, he advocated for federal quarantines of so-called dying Inuit cultures even as he debunked the notion of the "Blond Eskimo."[41] Not all expedition members had such prominent careers as bureaucrats, however. Vilhjalmur Stefansson maintained a grudge against certain scientists in Canada, including Anderson and Jenness, which effectively barred him from Canadian governmental science. Despite this, Stefansson was instrumental in shaping Arctic political affairs in Canada in the 1920s and again during World War II. Ernest de Koven Leffingwell, on the other hand, faded into obscurity in the 1920s, giving up his scientific career for a quieter life on his family's citrus ranch in California. As an older man, he reflected on how his research led to the creation of Naval Petroleum Reserve no. 4 on Alaska's north slope and the development of an oil empire there.[42]

Without downplaying or exaggerating it, expeditions transformed the Arctic—but not only because of scientists' efforts. Starting in 1918, Inuit from the Mackenzie Delta—which, by this point, included Inuit born in that area but also those born throughout coastal Alaska—converted the unwanted equipment of expeditions into their own implements of expansion. When expeditionary scientists departed the Arctic, they often jettisoned gear, to make room for scientific materials aboard the trading vessels they took home and to pay Inuit assistants for their services. Natkusiak, an Alaskan Inupiaq who worked on the Stefansson-Anderson Expedition and the Canadian Arctic Expedition, earned quite a prize with his payment, the schooner *North Star*. Outfitted with sails and decks, and complete with a cabin, this vessel allowed Natkusiak to visit trapping lands that were otherwise inaccessible by *umiak*—which required paddling and offered little shelter. Natkusiak took the *North Star* to Banks Island, north and east of the Mackenzie Delta, where he had witnessed abundant fox while traveling with Stefansson. Upon arrival, Natkusiak and his associates discovered parts of the *Mary Sachs* had been abandoned by the Canadian Arctic Expedition's Northern Party and used its supply of wood in constructing the beginnings of a new trade post. Now known as Sachs Harbor, this village has more than a nominal relationship with the history of science. Natkusiak became known as "Billy Banksland" for opening the lucrative fur grounds of Banks Island, which, in turn, solidified Herschel Island, Tuktoyaktuk, and Aklavik as trade entrepôts during the 1920s.[43]

Like Natkusiak, other western Arctic Inuit learned from expeditions about the biological productivity of lands on the edges of their travel routes—and capitalized upon them. Soon after Stefansson pronounced the discovery of

FIGURE 7. Inuit schooners at Herschel Island, circa 1928. Credit: Nunavut Archives / Marjory Robertson / N-1988-042: 0004.

the "Blond Eskimo" in 1912, several western Arctic Inuit entered the central Arctic in hopes of bringing that region into the growing trading economy. In 1916, whaler-turned-trader Charlie Klengenberg opened the first trading post on the Coppermine River. By that time, Klengenberg's Inuk wife, Qimnig, had given birth to six children. One of these, Patsy, became the manager of the Hudson's Bay Company's fur trading posts in the central Arctic in the 1930s after acting as a translator, guide, and assistant to the Southern Party of the Canadian Arctic Expedition. In the mid-1910s, several Inuit set up a small trading post near Walker Bay, in the heart of the Coronation Gulf. Elders from this region, belonging to what are now the Kanigiryuarmuit, remember these enterprising trappers and traders as "Westerners" (Mackenzie Delta Inuit), not the hunters common to the area. The Walker Bay post eventually became a permanent community in the Inuvialuit Settlement Region, first called Holman and later Ulukhaktok. It was settled by members of Klengenberg's family, Natkusiak, and a young multiracial man named Alex Stefansson—Vilhjalmur's son.[44]

For those looking back in time, schooners are symbols of Western Arctic Expeditions. They reflect the interventions of whalers, scientists, and governments, as well as the ways western Arctic Inuit responded to them (figure 7). Schooners extended the range over which Inuit trappers could find and sell furs in the 1920s, a decided advantage in trade. In 1917, no Inuit trapper owned such a vessel. By 1926, there were fourteen Inuit with them in the Mackenzie Delta area. None of these were crafted locally; all came from departing whalers and scientists. By 1935, more than fifty Inuit owned schooners—evidence of how important the technology became to Inuit livelihood. In these ways, Inuit who accompanied field-workers and received discarded gear became leaders of northern communities and economies in a subsequent generation. They translated knowledge and technology into sources of subsistence and identity, in the midst of rapid environmental change. In this case, the resilience of Inuit is as remarkable as the scientific project of trying to control the Arctic. Indeed, the two forces produced each other.[45]

In conceiving of the Arctic as threatened, scientists and governmental officials gave themselves power over northern life, even if they could not fathom Inuit reshaping the land on their own. Scientists, bureaucrats, and Native northerners would not fully realize this situation—and the disruptions it could create—until they met again in the Arctic. When botanist Alf Erling Porsild went to northern Alaska and the Mackenzie Delta at the end of the 1920s, he brought only four books with him. One was Stefansson's report of his 1908-12 expedition, *My Life with the Eskimo*, and the other a copy of Leffingwell's *The Canning River Region*. Yet neither account matched the landscape in front of him or the movements of Inuit on schooners across the western Arctic. The frontiers and debates that science had supposedly settled seemed reopened, even wilder than before. This was evidence to support a new colonial imperative after the Great War: tame the tundra.[46]

CHAPTER THREE

Wild: Taming the Tundra

Like a black mass of some fluid the herd slowly approached the edge of the plateau—began to flow down first slowly—a few deer at a time but soon gathering impetus and speed and ending in a wild rush. . . . It was a grand sight that I will never forget. . . . The drive is on its way to Canada.

ALF ERLING PORSILD, 1929

One was a Danish botanist raised in Greenland. The other was an American plant ecologist trained on the sheep lands of Utah. Alf Erling Porsild and Lawrence Palmer shared a purpose: transform the Arctic into reindeer country.[1]

They met at the depot in Anchorage in July, 1926. The Canadian Department of the Interior sent Porsild to learn about Palmer's work with the US Bureau of Biological Survey. Palmer had spent the previous seven years studying Alaska's vegetation and teaming with the Lomen Company of Nome to build a new industry in the territory. Reindeer, a species foreign to North America, but common in other parts of the circumpolar world, formed the center of this business. The biologists headed for Fairbanks so that the Canadian representative could see the foundation of the reindeer system. As they toured Palmer's field station and experimental farm, Palmer showed off his plant trials. Here, in fenced-off patches of lichens, sat the makings of another round of intervention in Arctic life.

In the period between the world wars, Palmer and Porsild generated new knowledge of plants, animals, and their relationships, and applied that knowledge in governmental and corporate-sponsored reindeer projects throughout the western Arctic. After his rendezvous with Palmer in Fairbanks, Porsild traveled north along the Beaufort Sea shoreline into Canada, where he surveyed thousands of square miles of tundra to find a suitable home for a Canadian herd. His official report of these travels, *Reindeer Grazing in Northwest Canada*, provided the basis for the Canadian government's choice of the Mackenzie Delta as a home for its Canadian Reindeer Project, the

Dominion's attempt to replicate the US Bureau of Biology's success. In 1929, Porsild returned to Alaska to meet Palmer again, this time to purchase three thousand reindeer from the Lomen Company and drive them to the delta. When the Danish botanist celebrated that "grand sight" in his field journal, he commemorated a historic moment. Reindeer ushered in a new era in the north.[2]

Ecologists effectively introduced reindeer and initiated systematic study and management of herding. Keeping these colonial projects viable over time proved to be a whole other animal, however. While six teams of Inuit became managers of herds in Canada after 1938, all of these operations collapsed by 1959. In that year, the Canadian government handed the project to private developers, having little to show for its million-dollar investment. Meanwhile, along the north slope of Alaska, reindeer herding followed a similar trajectory. The number of reindeer in Barrow—the largest town in the region, with a population over 1,200 Inuit—had dropped from an estimated 35,000 in 1935 to fewer than 5,000 by 1940. By 1952, domestic reindeer had disappeared from the entire Arctic slope of Alaska.[3]

On the surface, this turn of events seems easily disregarded as a harebrained colonial scheme. Such a dismissal, however, overlooks the nature of intervention in the Arctic, whether in the past or present. Federal agents rationalize their interference in northern life according to the scientific ideas available to them. Between the world wars, those ideas created a logic in which the reindeer could be a technology of administration and development. Rather than observe meteorological phenomena so whalers could secure their harvests, or bring samples of minerals, flora, and fauna back to museums, Porsild and Palmer reduced the Arctic to a few interactions between people and the environment and tinkered with these interactions, on site. Reindeer projects, then, were experimentalist in the sense of the scientific method as well as experiments for governmental scientists who had never before managed Arctic resources directly. Scientists explained that the tundra was wild. The rational approach—the natural thing to do—was to was tame it.

BRINGING REINDEER TO NORTH AMERICA: FROM CIVILIZING TO CONSERVING AND COMMODIFYING

Like their wild cousins, the caribou, reindeer adapted to the Arctic climate and their bodies provided nutrition, clothing, and tools to their herders. For missionaries in the 1800s, the reindeer's consumption of tundra vegetation promised a set of desirable conversions. The creature could turn nonarable hinterlands into productive grazing lands and primitive Inuit hunters into sophisticated animal ranchers. Sheldon Jackson brought a herd across the Bering Strait from Siberia to Port Clarence, Alaska, in 1892. This herd increased in size and spread throughout the Bering Strait and Beaufort Sea regions over the next twenty years. In Canada, Dr. Wilfred Grenfell spearheaded the introduction of reindeer to Newfoundland in 1908 with motivations that mirrored Jackson's in Alaska, but without the success.[4]

In Alaska, missionary and Inuit management of the animal yielded to governmental and corporate control, but not intentionally. An outbreak of Spanish influenza in the late 1910s and early 1920s devastated Native herders along the Alaska coast, crippling villagers' access to food and annihilating the sources of expert knowledge needed to manage reindeer. While the effect of the flu appeared to jeopardize the budding industry, it also created a business opportunity for capitalists nearby in Nome. The Lomen family, which had made its money in gold over the turn of the century, abandoned mineral extraction and began purchasing reindeer. By the mid-1920s, the Lomen Company was the largest private owner of livestock in Alaska with over fourteen thousand animals. Over the next three years, the Lomen Company forged relationships with high-profile businessmen and bureaucrats in the East to turn reindeer into an American industry (figure 8). Carl Lomen, one of the brothers heading up the family business, secured the backing of investors in New York City to purchase steamships, outfit them with storage compartments, and deliver reindeer meat from Nome to Seattle for distribution to restaurants in major cities throughout the country. Most importantly, though, Lomen recognized the value of science to postwar development. He curried favor with leading biologists, including E. W. Nelson, the chief of the US Bureau of Biological Survey. Nelson and Lomen agreed that the industry, and the vegetation upon which it was based, could be better managed with science. Nelson chose Lawrence Palmer, a biologist with the US Forest Service, to initiate a research program on grazing ecology in Alaska.[5]

The reindeer's viability as a technology of colonial development hinged on power brokers like Lomen and Nelson seizing the moment. Previous investments in minerals, for instance, kept the Lomens out of the reindeer game until the flu opened a door. Similar deliberations characterized the situation in Canada. As Palmer set up in Alaska, members of the Canadian Arctic Expedition of 1913–18 returned to Ottawa to champion reindeer as part of a strategy for managing the Arctic. Having passed through Nome on their

FIGURE 8. One of the herds operated by the Lomen Brothers gathers in the vicinity of Kotzebue, Alaska, in 1915. The Lomen Brothers used photographs like this to promote the reindeer industry among bureaucrats and boosters in the metropolitan United States. Credit: P240-215, Alaska State Library, George A. Parks Photo Collection.

way in and out of the north, these scientists had witnessed the growth of the Lomen Company's reindeer operations and observed Inuit-managed herds on the Beaufort Sea coast. Their missions in the field, however, precluded them from performing careful analyses of reindeer biology or the economics of herding. In turn, while these scientists spent some of their time in the 1920s drafting reports of the phenomena and materials they did collect, they also lobbied the Canadian government for new studies of reindeer. Their efforts persuaded the government to convene a Royal Commission to "investigate the possibilities of the reindeer and musk-ox industries in the Arctic and sub-Arctic regions of Canada." In a series of meetings held in 1919, the commission heard from whalers, missionaries, northern police, elected officials, and eight of the scientists from the Canadian Arctic Expedition. Just as Carl Lomen built interest for his Alaskan enterprise, these researchers promoted reindeer country as a potential meat-producing factory for Canada.[6]

Scientists also introduced a conservationist sentiment within the elaboration of reindeer industries. They worried about the possibility of caribou extinction, given the reverberating impact of commercial whaling on the caribou populations on the Yukon Coast and Mackenzie Delta—something they documented while on Western Arctic Expeditions. Scientists invoked the historical example of bison on the Great Plains to finger Inuit hunters as a great risk to the caribou's continued survival. As historian John Sandlos has thoughtfully argued, Canadian researchers framed introducing reindeer as means of commodifying northern prairies and of protecting native wildlife. By encouraging Inuit to take up herding, they could prevent them from hunting—a seeming win-win for wildlife management and economic development in the north.[7]

WILDLIFE AND TAME-LIFE

While game conservation and reindeer introductions extended colonial jurisdiction in the far north, the two impulses were different beasts. Taming the tundra required unique relationships with Inuit that could be sidestepped in regulating caribou hunting and creating national parks. In turn, engaging Inuit changed how scientists and bureaucrats imagined northern development.

The distinctions are brought into focus when attending to the legal relationship between governmental agents and Inuit in the early twentieth century. In Alaska, the US Bureau of Education teacher acted as de facto government. Because territorial officials attached reindeer herds to schools, enrolling Natives as herders also enrolled them in the federal bureaucracy. Such oversight had not yet been possible in Arctic Canada. Of course, Royal Canadian Mounted Police and missionaries imposed legal and moral codes at trading outposts in Inuit territory. Still, Inuit had never dissolved rights to the land, and thus federal agents in the United States and Canada were eager to find some apparatus to develop both Inuit and Arctic resources. Indeed, in the contemporary case of muskox conservation, consultants to the Canadian Advisory Board on Wildlife Protection suggested the federal government practice diplomacy with Inuit to enlist northerners in the project of protecting nature. In terms of turning Inuit into subjects of the state, this seemed a more effective alternative to doubling-down on hunting regulations that could not be adequately enforced.[8]

Accounts from Inuit also demonstrate reindeer as a means of law. According to Inuvialuk Randall Pokiak, Inuit living in the Mackenzie Delta and along the Arctic coast in the early 1900s were troubled by the recent influx of Alaskan Inupiat into the area, as they deemed these foreign Natives responsible for a recent decline in caribou populations. Alaskan Inupiat had traveled eastward since the 1880s, first with commercial whalers who had overharvested caribou in the Bering Strait and north slope regions, and later to avoid the epidemic of Spanish influenza after 1918. Calling on a local shaman, leaders of the Mackenzie Delta Inuit hoped to alter the migration patterns of caribou to force the Inupiat back home. The caribou did go away, but did not return. Native Mackenzie Delta Inuit thus became amenable to alternative methods of procuring food. When Canadian government agents approached them in the 1920s to sign a treaty, Inuit leaders asked for the delivery of reindeer from Alaska, having "heard stories from Inupiat that reindeer had the same diet as caribou." Oral histories indicate that one Inuk, Mangilaluk, negotiated with the government, suggesting to treaty officers that, if they brought reindeer, "they would think about signing an agreement." In both Alaska and northern Canada reindeer created possibilities for domesticating Inuit and the Arctic, whether through religion, education, or law.[9]

Reindeer herding, and thus Inuit, also became key mechanisms in plans for Arctic economic development. After the Canadian Arctic Expedition of 1913–18 returned with the promise of extensive copper deposits in the Coronation Gulf, federal agents withdrew these from private staking and then sat on them, foregoing development because other national resources were more accessible. Scientists argued that, in order to capitalize upon Arctic resources in the future, a local food source would need to be established now, since populations of migrating caribou had been decimated. Many southern Canadians believed that white men were unlikely to want to live in the north. Reindeer and Inuit offered solutions to these problems. Inuit could be responsible for maintaining reindeer herds, the meat from which could be shipped to the Coronation Gulf, reducing overhead costs for privately or federally sponsored mining.[10]

Drawing connections among labor needs, environmental changes, and the possibilities of reindeer and mineral economies, promoters of northern development articulated Inuit as a valuable asset. This was a paradigm shift from earlier conceptions of northern peoples as dangerous, primitive, and useless in terms of staking legal claims to the Arctic. Anthropologist Diamond Jenness, who had invalidated theories of the "Blond Eskimo" after his service on the Canadian Arctic Expedition, distilled the situation for his audience at a 1923 lecture at the Victoria Memorial Museum. "Unless we use the Eskimos," he argued, "we can never develop the Northland." This reorientation toward Inuit paralleled shifts within Canada's Department of External Affairs, which abandoned a politics of possession after the Great War in favor of a foreign policy that emphasized human occupation and resource extraction. Reindeer thus helped southerners navigate a postwar geopolitical landscape where the only frontiers acceptable for exploitation were internal. A wild Arctic—one filled with people, but in need of improvement to reach its full potential—thus became a compelling rhetoric of progress across the southern United States and Canada.[11]

CALLING ECOLOGY AND DEVELOPMENT INTO BEING

Lawrence Palmer had been sent north by Edward Nelson, the chief of the Bureau of Biological Survey, to carry out research in service of the government and the Lomen Company. But it was not as clear in Canada who would lead the Dominion's reindeer experiment. The 1919 Royal Commission on Muskox and Reindeer, for example, called thirty-five witnesses to testify on the opportunities and obstacles facing a reindeer industry in Canada. That no trained botanist had spent enough time in reindeer country to give evidence before the commission did not appear to be an issue, though it eventually became one. The hiring of Danish botanists to lead the Canadian Reindeer Project thus provides perspective on the nature of scientific expertise. Far from being obvious or inevitable, the role of scientists in mediating environmental change in the Arctic is tenuous. It is called into being at particular moments by particular governmental priorities.

Consider the testimony from Royal Commission members on issues suspected to be indicators of failure or success of reindeer introductions. Expeditionary scientists detailed the extent and distribution of vegetation in certain geographical districts; the presence or former presence of caribou—which was assumed to denote the potential for reindeer; and prevailing winds (to account for a troublesome pest, the mosquito). Given this data and northern Canada's similarities with Alaska, Siberia, and northern Europe—all areas with thriving reindeer industries—witnesses believed animal husbandry would finally capitalize upon "vast tracts of country that are not utilized." The commission outlined vast swaths of the north as Canadian reindeer country, including several islands in Hudson Bay, the entire Ungava and Mackenzie Districts, the interior of the Yukon, and the Arctic coast from the international boundary to Kent Peninsula. Yet in the context of introducing an animal into these environments, there were also concerns with existing stores of knowledge.

Commissioners admitted there was much "conflicting evidence" about whether Inuit would take to herding, how reindeer managed pests, and how much time plants needed to recover after grazing. Commissioners underlined the importance of continued governmental presence "to remove the elements of doubt and uncertainty, and so tend to encourage private enterprise and investment." This could be accomplished through "careful study" of individual localities, so as to "utilize to the best possible advantage, as means of control, any suitable valleys or other special topographical features, which may be available." Participants agreed that the Canadian government should lead the initial reindeer trials, beginning with a small, manageable herd, working out any kinks in logistics, and paving the way for future investment by private groups. In these ways, state plans for reindeer relegated the significance of previous studies, and created new gaps in understanding, yet did not challenge science as a way of colonial thinking.[12]

Despite this faith in governmental and scientific development, the first attempt to cash in on reindeer in Canada after the Royal Commission came through private enterprise. Resigning from the body, anthropologist-explorer Vilhjalmur Stefansson introduced reindeer to Baffin Island in 1921, in conjunction with a new subsidiary of the Hudson's Bay Company, the Hudson's Bay Reindeer Company. Confirming concerns from governmental scientists, this partnership failed to yield administrative capacity, profits, or cultural improvement. In 1921 the herd reached Baffin Island, and by 1927 most of the reindeer had died or disappeared, prompting the government to cancel the Company's grazing permit. Newspapers across the United States and Canada covered this story. Stefansson interpreted their scathing remarks as damaging his reputation as an expert on northern matters. This debacle eventually catalyzed the hiring of botanist Alf Erling Porsild, who went on to direct Canada's experiments with reindeer in the Arctic until the mid-1930s.[13]

Stefansson was known for his contentious nature, but this case was as much about changing requirements for knowledge about northern development as his knack for the spotlight. Before 1921 and the formation of the Hudson's Bay Reindeer Company, members of the Canadian Department of the Interior relied on substantial northern experience—like the kind Stefansson had amassed in his ten years in the Arctic, or that embodied by the witnesses to the Royal Commission on Muskox and Reindeer—rather than academic training when appraising the scientific needs of Arctic administration. Those with both academic expertise and northern experience, like many of the scientists who had served on Western Arctic Expeditions and were thus invited to the commission, seemed especially useful sources. In early 1926, though, the Department of the Interior refined their interest to a particular type of knowledge and know-how—applied botany.[14]

The shifting definitions of Arctic authority materialized in correspondence among Canadian bureaucrats trying to decide on a suitable manager for the Canadian Reindeer Project. In January 1926, the head of the Northwest Territories and Yukon Branch, O. S. Finnie, could see little value in a botanist. He wrote the deputy minister of the interior, making a plea for a man with practical skills to lead a governmental reindeer project. "I do not think the qualifications as a Botanist is sufficient," Finnie noted, because "I believe we would get better results if we could get a practical reindeer man who knows the kind of feed that the reindeer live on, and one who is a good traveller and could go through the country and size up the situation accurately and quickly." By 1927, though, Finnie expressed a firm commitment to applied botanical science as a way of knowing and managing reindeer. When the Dominion Reindeer Company of Vancouver inquired in 1927 about leasing land in the Northwest Territories, Finnie responded with caution. He was unable to recommend any location "until the different districts in the North West Territories had been thoroughly cruised with a view to determining their value as feeding grounds for the reindeer." Finnie granted a lease to the Dominion Reindeer Company in the eastern Arctic in 1928, contingent on a scientific survey of the region. They never completed such a survey, prompting the government to terminate the lease in 1931. Finnie asserted that the recent history with the Hudson's Bay Company had "served as a lesson" for governmental managers of reindeer experiments.[15]

Finnie and other reindeer enthusiasts in Canada were also convinced of the value of applied botany by their counterparts in Alaska. In March 1926, the high-ranking Canadian official W. W. Cory visited Washington to consult with US officials on best practices for a Canadian reindeer industry. While there, Cory met with Dr. E. W. Nelson, the chief of the US Bureau of Biological Survey. Nelson impressed upon Cory that a single man could not handle the duties of getting the Canadian Reindeer Project off the ground. These duties involved conducting surveys and experiments on reindeer, including their principal movements, feeding habits, and major predators, pests, and diseases. They also required surveying Canada for suitable forage and building the systems of reindeer management, like the supervisory hierarchies, corrals, and storage facilities needed to round up, slaughter, and process reindeer. Most importantly, though, Nelson advocated for trained ecologists to fill these roles. The importance of ecological studies of reindeer feed had been made clear by the bureau's grazing scientist, Lawrence Palmer, who had studied under one of the United States' leading ecologists, Frederic Clements. Nelson attributed the steady growth of reindeer populations in Alaska to Palmer's ability to translate his field experiments to the management of grazing lands. In 1901, one thousand animals roamed the coasts of Alaska; in the mid-1920s, that number had exploded to over two hundred thousand. Palmer had also argued that, when his research was fully applied, reindeer country in Alaska could support three million livestock. Nelson suggested that the Canadian Department of the Interior hire two botanists and have them apprentice with Palmer for six months, learning reindeer ecology and the reindeer business. Cory relayed this news to Finnie, and with both men sold on the model of the Alaskan industry, they began to see botanical expertise and ecological research methods as fundamental to modern reindeer management.[16]

Governmental priorities for the Arctic in the 1920s, along with the political economy of the moment, called ecology into being as a form of colonial authority. This fact materialized in the hiring of Alf Erling Porsild and his brother, Robert Porsild, to lead the Canadian Reindeer Project. They grew up in the shadow of an Arctic research station in Greenland, within a transient community that offered useful training in northern botany and Arctic travel. Their abilities to speak an Inuit language and thrive in conditions many southerners considered harsh also met the expectations of the Canadian Department of the Interior and the US Bureau of Biological Survey. If Finnie had been initially resistant to the value of a botanist, other northern promoters bristled against the shifting domains of credibility surrounding reindeer too. The Porsild brothers' lack of practical work with reindeer was not lost on Vilhjalmur Stefansson, who lobbied his peers to reconsider their hiring. These deficiencies did not bother Finnie, Cory, and Nelson, who came to believe in the Porsilds' talents and skills, and were convinced that time in Alaska spent gaining hands-on experience with reindeer and grazing ecology would fill the remaining gaps. While his brother, Robert, eventually left the reindeer business, US and Canadian bureaucrats and scientists soon identified Alf Erling Porsild as a leading expert on Arctic vegetation and reindeer.[17]

LICHENS AND LEGIBILITY

In May 1926, the Porsilds and Palmer left the train station at Anchorage, embarking on their Arctic studies. They carried a note from the director of the US Bureau of Biological Survey to serve as a set of instructions for the Alaskan ecologist. Palmer was to offer the Porsilds his "fund of information" on reindeer. This fund had been generated by his quadrat studies on tundra regrowth and carrying capacity at the Fairbanks experimental station and his collaboration with the US Bureau of Animal Industry on the nutritive quality of various types of forage. E. W. Nelson also recommended that the Porsild brothers learn the practical workings of the herds, trying their hands at corralling, capturing, marking, castrating, and branding.[18]

Discerning the Porsilds' apprenticeship with Palmer is crucial to understanding the fate of reindeer in the Arctic and the consequences of taming the tundra, both intended and unintended. This partnership guided the Porsilds in siting the Canadian Reindeer Project and crafting its inner workings. Palmer emphasized the importance of a particular kind of knowledge in first selecting and subsequently managing a reindeer grazing area. The Porsild brothers, Nelson wrote, "should be taught as much as possible concerning the forage plants used by these animals, with a special view to the differences between the summer and winter forage and the need of safeguarding the winter forage areas from use in summer," so that the range, and governmental power, could be perpetuated. Recognizing and protecting forage were foundational to managerial decisions in Arctic animal husbandry economies. As such, these twinned convictions have been inscribed onto the physical and social landscapes of the north slope of Alaska and the Mackenzie Delta, even if governmental reindeer projects died off after 1950.[19]

To understand how this could be so, we must first gather the details of what the Porsilds learned in Alaska and, thus, become familiar with the work of Lawrence Palmer. Palmer studied at the University of Nebraska between 1911 and 1915 before becoming a grazing assistant with the US Forest Service. Hired in 1919 as an assistant biologist for the Bureau of Biological Survey, Palmer considered himself a botanist, biologist, ecologist, and range manager—suggestive of the multiple relations among ecological plant studies, agricultural development, and state control at the time. He applied his knowledge of grazing relationships in the American West to the study of reindeer. His first five years in Alaska were taken up with reconnaissance surveys of the herds along Alaska's meandering coastline. These surveys supplied Palmer with a sense of the reindeer industry in Alaska, and the seasonal movements of people and animals across the land. As with range management in the West, Palmer concluded that the bases of the industry were the major species of plants that provided nutrition for reindeer. He arrived at this conclusion after careful study of these plants in the field and at the experimental station in Fairbanks.[20]

Nelson's instructions to impress upon the Porsilds the significance of winter forage likely did not surprise Palmer. After all, it was the Fairbanks biologist who had first articulated the significance of this component of the reindeer industry. Palmer developed an elaborate system of experimental pastures and quadrat studies in Fairbanks. These he explored with several lines of research, including the conditions governing forage and range management, the various relations of lichens to grazing, the relative carrying capacity of lichen and nonlichen ranges, and the methods of feeding and their effects. Palmer parceled out eleven pastures, each with slightly different vegetation based on its position on the slope of the hill on which the farm sat. He brought reindeer to graze within these pastures, learning about how the animals ate, what plants they selected in different seasons, how they dealt with snow, and how the plants responded in spring. He established quadrats within these different pastures and performed his own tests, cutting plants and picking them by hand. These experiments convinced Palmer that winter forage, comprised mostly of the genus *Cladonia*, was the keystone in the architecture of a modern, successful reindeer industry. Palmer was certain that the study of lichens would also open an entire field of inquiry for Alaskan and broader scientific communities interested in the ecology of plant succession, beyond supplying the local industry with valuable data.[21]

By 1926, Palmer had made a case for organizing the entire industry around *Cladonia*. He noticed that winter ranges were patchier than summer ranges. Winter resources, then, had to be protected—especially given the observation that reindeer bunched up in colder temperatures, potentially overgrazing their food source. A closer look at the nutritive quality of winter forage plants and their reaction in quadrat studies to mowing, picking, and feeding showed surprising results. After only a few years of observations, Palmer noticed that it might take winter lichen ranges ten to fifteen years to "come back to a normal height growth of four to five inches." Thus proper management of the winter range presented Palmer with "an exceptionally important problem." After allotting time for recovery of the winter range, the ecologist asserted that carrying capacity must be on the order of 40 to 60 acres per head. Extrapolating to the available land in Alaska suitable for grazing, he estimated the territory could support three million reindeer, three times as many the fully stocked industry had in 1926.[22]

Palmer's concerns and visions for the reindeer industry must have been colored by his experiences with the Forest Service in Utah. He was a biologist in Ogden, part of Region 4, which boasted more acreage under federal supervision than any other part of the country. It also boasted more cows, sheep, ranchers, and problems with soil erosion. Working at the Great Basin research station between 1915 and 1919, Palmer had been exposed to the Forest Service's goals for watershed protection and the scientific studies that helped meet those goals. In Region 4, managers balanced multiple uses of the land—recreation and grazing, for example—with perpetuating natural resources into the future. They based their decisions on information gathered from "period studies," or those that determined the "season" for grazing by documenting the life cycles of plants and organizing the movements of animals to prevent overgrazing of new, young plants. Region 4 led the United States in these types of studies and helped make common the use of quadrats as research tools. Palmer's work in Utah also helped solidify the broad concept of "carrying capacity," a measurement of the maximum amount of grazers the land could support without becoming exhausted. When he made recommendations for the future of Alaska's reindeer, Palmer synthesized his personal experiences in the West and the North, as well as a diverse set of ecological relationships among plants, soil, animals, and humans.[23]

Palmer's conclusions about *Cladonia* and carrying capacity fit into a larger Bureau of Biological Survey scheme of modernizing the Arctic. Palmer lamented that reindeer handling in Alaska suffered "from lack of application of improved modern methods." What he meant was modern science, and more specifically, the concept of rotational grazing. This was another concept Palmer imported to Alaska through the Department of Agriculture from range science in the West. In theory, this approach made maximum use of forage available by moving herds between a series of summer and winter pastures and preventing overgrazing by allowing some tracts of land to go fallow each year. In order to make this kind of grazing possible, Palmer noted, the industry's management and infrastructure would need to be overhauled. The territory must be divided into grazing units, which could be delimited by "natural monuments" such as divides between drainages. Corrals should be constructed to facilitate roundups and slaughter. Permanent winter cabins should be built to ease herd management in winter, the most important phase for the protection of *Cladonia*. But most importantly, rotational grazing depended on open herding, where animals were free to select food on their own. This approach contrasted with the tradition of close herding, practiced by Sami, where herders and animals stayed together as they moved over the land. Both Palmer and Nelson agreed that rotational grazing replaced the "crude methods of the original herders" and instilled in the industry "definite scientific investigations [and] oversight."[24]

The modern methods advocated by Palmer and the Bureau of Biological Survey dovetailed with corporate plans for Alaska. Starting in Nome and working their way outward across the Seward Peninsula and nearby districts, the Lomens began buying reindeer and creating a vertically organized corporation, where they owned herds, corrals and processing facilities, ice shelters for meat storage, and ships with insulated hulls for refrigerated transport to Seattle. Open herding fit within this business model, as it was limited only by the size of the grazing allotment, not the amount of labor needed to move with the herd, as was the case with close herding. The Bureau of Indian Affairs also advocated open herding at the time, given the outbreak of influenza across Alaska, which had killed many of the experienced herders. Despite immense growth of reindeer villages in the territory after the collapse of whaling and the gold rush, the reindeer industry found itself in desperate need of capital, both human and financial, to stay alive. Open herding seemed to provide a safety net for the Inuit families in the industry, as it allowed a small number of inexperienced herders to manage a herd together, sharing knowledge and labor.[25]

Open herding was not without its problems, however. It encouraged straying of animals and attacks from predators that would not be scared off by the presence of herders. Some Bureau of Indian Affairs and Bureau of Education staff also suspected that open herding was not creating the deep knowledge and respect of nature that close herding had. They wondered if open herding provided fewer cultural benefits for so-called savage Inuit hunters. The thorniest issue, however, was mixing. When Lomen herds mixed with local Inuit herds—which they inevitably did in areas without fences—herders often corralled large groups of the animals so they could be separated. In almost all cases, some reindeer were found to have no brand to indicate the owner. These unmarked animals were distributed to owners based on a percentage scheme, itself derived from the size of the herds, with the larger herd receiving more of the share of the unmarked animals. This practice likely encouraged the Lomens not to worry about mixing, as they stood to gain nearly as many animals from a well-supervised herd as they were a herd subject to straying.[26]

Palmer's research provided the foundation for these interactions among herders. He had worked closely with the Lomen Company, the Bureau of Indian Affairs, and the Alaska Game Commission to create the procedures around branding animals. He advocated for distributing unmarked animals in mixed herds, thereby aiding the growth of the industry and the company's growing influence within it. His stated aim was to unite "scattered herds into

FIGURE 9. Lawrence Palmer converts the Arctic landscape into the terms of reindeer ecology. Map by Lawrence Palmer, *Progress of Reindeer Grazing Investigations in Alaska*, Bulletin no. 1423 (Washington, DC: US Department of Agriculture, 1927), 2.

large herds, each under white supervision with natives as herders." The tight association here between industry, science, and social control would later form the center of a debate over reindeer ownership in Alaska, resulting in a 1937 law that disallowed non-Native ownership. But until that point, the union of governmental and commercial interests largely buried complaints from Alaskans about the use of science in the service of corporate greed and colonial intervention.[27]

As we consider the relationship of Porsild and Palmer—and the connections among science, the state, and reindeer—we must remember the concepts of winter forage and carrying capacity were components of a management regime. This regime made room for the expertise of scientists to guide the activities of Sami herders and Native apprentices. To visualize the linkage between scientific knowledge, state supervision, and the reindeer industry, consider the map Palmer presented to his readers in his 1926 Department of Agriculture publication (figure 9). Through reference to "Eskimo allotments," "White man allotments," and "Unoccupied grazing areas," Palmer argued for the merit of a rational, scientific manager to preside over people and nature in the north. This was a person who could hold together the particular nutritive value of *Cladonia* and the landscape mosaic of topography, vegetation, and climate to direct the right herders to the right places at the right times. Winter forage and carrying capacity were thus mechanisms for legibility, the capacity of governments to represent the resources of particular territories so as to exploit them. Palmer's ecology provides a telling example of what political scientist James C. Scott calls "the radical reorganization and simplification of flora to meet man's goals."[28]

HOME ON THE RANGE

When the Porsild brothers were given orders to learn what Palmer had to teach them about reindeer, a passage was opened between the Canadian Reindeer Project and scientific ideas emerging from the Fairbanks station. Palmer walked the brothers through the practices of marking, corralling, and butchering and shared "all his reindeer files" with Alf Erling Porsild. The bureau biologist also conveyed his views about the advantages of open herding and, by association, the superiority of "modern" methods for handling reindeer over Native Alaskan and Sami ways of knowing the animal. The Porsilds left Nome, Alaska, in December 1926, completing a trek to the Mackenzie Delta to test a possible route for the delivery of the herd to Canada. Upon arriving in Aklavik, Alf Erling Porsild wrote to O. S. Finnie, the director of the Northwest Territories and Yukon Branch, to proclaim the reconnaissance mission with Palmer a success. Noting the plant cover in the Mackenzie Delta flats, Porsild characterized them as one of many tundra "types," which "entirely conform[ed] with similar deltas of Buckland, Kubuk, or Noataq in Alaska." To tame the Arctic, it seems, one first needed to learn how to see the state of nature.[29]

Between 1926 and 1931, Alf Erling Porsild visited Alaska twice and scoured the Canadian north for a home for reindeer. Ultimately, Porsild recommended that only two districts, the Mackenzie Delta and the Dease River valley, were suitable for governmental reindeer. When viewed together, these activities and the reports Porsild wrote about them reveal how Palmer's regime of reindeer configured Porsild's observations, his conclusions, and the construction of science and colonialism in the Arctic.

Between April and August 1927, Alf Erling Porsild and his brother completed a survey of the "Husky Lakes" region between present-day Inuvik and Tuktoyaktuk. With excitement, he pictured the region with Palmer's lichen ecology in mind. "*Cladonia* . . . probably covers more ground than all the rest together," he wrote. "This lake would be ideal location for winter reindeer camp." Based on his observations of forage in the area, Porsild estimated the country could support up to a quarter of a million reindeer. The carrying capacity of the Husky Lakes region did not dwarf other areas Finnie asked the Porsilds to study. In 1928, the brothers inspected the valley of the Dease River, which extends northeast of Great Bear Lake, and the "northern plains" running south and west of the lake. Porsild described the region as a "natural grazing unit" as it was "closed in from all sides," and afforded abundant vegetation. He increased the number of acres that should be allotted per reindeer there, though. The units presented a different "tundra type" than the Mackenzie Delta, and Porsild found it difficult to estimate grazing potential in this "unmapped country." Still, the botanist suggested that the twenty-five million acres of the Great Bear Lake basin could support a total of three hundred thousand reindeer.[30]

Alf Erling Porsild's evaluations of the Arctic make clear how ecological awareness helped site the Canadian Reindeer Project. Consider the problems he identified in the Great Bear Lake basin. The southern shores of the lake were "too heavily timbered to make herding and control of tame reindeer practicable." But more importantly, both the Dease Valley and the northern plains units offered little protection from mosquitoes, the ubiquitous, yet temporary, pest of reindeer industries. Porsild peppered his diaries with comments about how annoying mosquitoes could be, as well as how troublesome they were to effective reindeer management. When visiting Palmer in 1929, Porsild learned that nearly fifty of Palmer's stock at Fairbanks had been killed by mosquitoes in the previous year. In 1936, Porsild spelled out the consequences of mosquitoes for the potential expansion of reindeer industries. "Nowhere in the area under consideration are the hills high enough to permit reindeer to escape flies during the summer," he concluded. For this reason, Porsild surmised, reindeer ranching would be "limited to the seacoast and adjacent hinterland."[31]

Such characterizations of the nature of northern development stood in stark contrast to previous instances of science and intervention in a dangerous or threatened Arctic. Reindeer were not suitable across the far north, but rather fit a niche in certain ecosystems. Ecologists encouraged a more sophisticated knowledge about the north within federal governments, which tailored administrative action to the particularities of geography, climate, topography, and biology. Representations and interventions, though shifting in scale or character, were always intertwined.

The relations of ecological knowledge and northern colonialism found their clearest expression in descriptions of vegetation in the Mackenzie Delta, the eventual home for Canadian reindeer. With rotational grazing in mind, Porsild admired the patchwork of tundra plants evident in the Mackenzie Delta and Arctic coast. He employed Palmer's models of tundra types and the Alaskan scientist's ecological counting methods to determine the exact proportion, distribution, and nutritive values of sedges, grasses, and lichens there. Ranking the "Husky Lakes" region as the best winter grazing land in Canada, Porsild commented on its "high percentage of palatable species," and the possibility for its "maximum development." In one turn of phrase, Porsild even pictured reindeer in this winter pasture "put[ting] on their back fat," directly linking the growth of plants with the growth of a northern reindeer industry.[32]

THE SOCIAL ORGANIZATION OF THE REINDEER PROJECT

Alf Erling Porsild did not apply his new knowledge about the north only to site selection. He also parlayed his growing expertise on circumpolar affairs to the institutionalization of the Reindeer Project over the 1930s. The continued investment in the Canadian Reindeer Project during this period is remarkable, given the retrenchment of the civil service in Canada and the reorganization of northern bureaucracies after the Great Depression. Clearly, Arctic science and Arctic interventions were top national priorities, and Porsild was ideally suited to execute them both.[33]

Porsild helped build out the Reindeer Project, in both a literal sense and a legal sense. In 1931, he traveled to Kautekeino, Norway, contracting three Sami families to relocate to the Mackenzie Delta to train Inuit in reindeer herding. Such contracts had become commonplace during the early 1900s in the region, as reindeer herdsmen confronted in-migrating farmers and attempted to resolve disputes over land. Sami herders would tend animals for settlers in exchange for assistance with lodging and other benefits. These arrangements seem to have formed the basis for the terms of agreement between the Canadian government and the herders recruited to travel to the Mackenzie Delta.[34] In 1932, Porsild returned to the east channel of the Mackenzie River to help construct the headquarters of field operations, Reindeer Station, which offered housing to Sami instructors and Inuit apprentices. In 1933, following Porsild's recommendations, the Canadian government withdrew a 6,000-acre tract for a Reindeer Grazing Reserve adjacent to Reindeer Station and established federal ordinances to protect reindeer as a national resource. These actions confirmed that nonherding Inuit could not hunt reindeer or otherwise enter the reserve without a permit, and those enrolled in the project could never become outright owners of reindeer, even if they graduated from apprenticeships. Shortly after, the Interdepartmental Reindeer Committee, having formed to consult the Department of Mines and Resources staff on the best practices for the reindeer industry, nominated Porsild to become the first superintendent of the Canadian Reindeer Project. In a matter of a few years, he had erected the scientific, physical, and political infrastructure of a new industry, from the ground up.[35]

More astonishing, however, was Porsild's ideas for the social organization of the project. While crossing from Barrow, Alaska, to Aklavik in the winter of 1926/27, Porsild met Tarpoq, an Inupiaq man and owner of a reindeer herd. Porsild found that Tarpoq was "a good reindeer man under the custody of a white man," but when he had been left unsupervised by the US Bureau of Education, he started to "neglect his herd when his increase and profits is [sic] not up to his expectations." In his journals and his reports to the Department of the Interior, he translated these experiences into evidence for a hierarchical regime of supervision over indigenous peoples. Porsild set up two new positions at Reindeer Station—a head foreman and an assistant supervisor. Their duties included controlling wolf predation through hunting and filing reports to Ottawa about reindeer losses. Leaving the actual herding of the reindeer to Sami, the foreman was also free to keep a closer eye on Inuit. He was responsible for recruiting apprentices from nearby settlements, issuing rations to all herders, and recalling herds from those Inuit who proved incapable. The qualifications of the foremen hired for the job are telling. All were white men who had spent time in the north, and many had extensive backgrounds with the military and the mounted police. In the minds of Porsild and other Canadian bureaucrats, someone with this pedigree could inculcate discipline within Inuit and administer a chain of command between the Delta and Ottawa.[36]

While Palmer had indoctrinated such racist perspectives within Porsild, ideas about Inuit culture permeating Canadian anthropology also animated the botanist's recommendations for governmental oversight. On the Canadian side of the Beaufort Sea coast, Porsild noted that the Inuit there had "too much easy cash" and had not yet learned the value of caring for their possessions. Moreover, he found that in the region between the international boundary and the Mackenzie Delta, Inuit had given up their customary seal hunt in favor of trapping fox. The fur economy, he found, afforded Inuit enough money to buy dog food (rather than hunt seal for it) and purchase other goods of interest, like flour, tobacco, ammunition, and rifles. In comparison to Sami, who rarely required a day off work and stuck to contractual obligations, Porsild noted that Inuit were lazy and capricious. He suggested that recruitment to herding focus on teen boys, since they were impressionable, less interested in economic pursuits, and unencumbered by family responsibilities. Like many anthropologists at the time, Porsild considered the adult population of Inuit of the western Arctic out of touch with a "traditional" life on the land and thus in need of guidance. He could not comprehend how a previous era of colonialism had given rise to the Inuit-led fur economy, let alone the ways Inuit appropriated new technologies and commercial exchanges into their own definitions of culture. Only in such a twisted colonial logic could reindeer herding—a livelihood foreign to Inuit—become a rational means of domesticating the north.[37]

ECONOMIC VOLATILITY AND INUIT RESILIENCE

By the mid 1930s, reindeer inhabited a landscape that looked quite different from that Porsild and Palmer had initially surveyed a decade earlier (figure 10). In the Mackenzie Delta, Sami herder Mikkel Pulk and Inuit apprentices pushed the main herd in spring from its winter range to the coastal area, where fawning commenced in early April and lasted until June. Reindeer were driven to Richards Island in the summer, where consistent sea breezes dispersed mosquitoes. Before the annual roundup, herders caught fish, harvested whale, and prepared this meat for the long winter. Reindeer foremen, hired through the Department of Mines and Resources, maintained files on nascent Native-run herds, sent paperwork to Ottawa to be reviewed by the Interdepartmental Reindeer Committee, and perused studies about herd management from scientists in Alaska. These supervisors could report a doubling in the Canadian herd between 1935 and 1938, despite slaughter-

FIGURE 10. Reindeer herders and their supervisor at a camp in the Reindeer Grazing Reserve. Credit: NWT Archives / Hadwen / N-1979-567: 0078.

ing more than three hundred deer each year to supply food and clothing for herders and two hospitals in Aklavik. That these movements of foremen, Sami herders, Inuit apprentices, reindeer, and scientists had become routine belies the transformations in scientific understanding and governmental capacity that had taken place in North America since the end of the Great War. It also belies similar transformations to come. By World War II, these same governmental reindeer projects had crumbled.[38]

To follow their demise, we must set aside the research efforts of Palmer and Porsild and take stock of global markets at the time. When the Great Depression hit, it crippled the legs upon which governmental reindeer programs stood. In Alaska, the crash in stock markets jeopardized the Lomen Company's solvency as exports of reindeer carcasses to the Lower 48 fell from the tens of thousands to just over six hundred in 1931. Adding insult to their injury, the success enjoyed by the Lomens in the 1920s now appeared as a threat to the US cattle industry. Cattlemen and their representatives in Washington worried about the inroads reindeer had made in beef markets. In response to these conditions and complaints, the US government called a series of hearings in Washington to evaluate Alaskan reindeer. Bureau of Indian Affairs commissioner John Collier advocated a government takeover of reindeer to create village governments and encourage Alaska Natives to develop their own economic activities in the region. Such a move required bureaucrats to buy out the Lomen herds and redistribute them to Alaska Natives. In 1936, Congress passed the Alaskan Reorganization Act, which put these plans in motion. The next year, Congress passed the Reindeer Act to authorize $700,000 to be spent moving reindeer from the Lomen Company to the hands of Native owners.[39]

These changes in Washington and Nome, Alaska, shaped reindeer herding on the Arctic slope. The Reorganization Act required that Native villages form joint-stock corporations to manage reindeer, which rendered obsolete the support of apprenticeships through Bureau of Education schools. This business model did not work well with Inuit herders on the north slope, since their labors supported people who did far less work, but still were paid through shares of meat and hides. In response, fewer Inuit entered the herding business and left more herds unsupervised. While straying, mixing, and predation remained constant problems, herders in northern Alaska also took to killing animals for nonfood purposes, such as baiting fox traps. Such Inuit responses are useful reminders about the nature of colonial power at the time. Governments and scientists could influence how each other viewed reindeer and the Arctic, establish jurisdiction in a legal sense, and even manipulate the region's physical environments in the name of national development. Yet Inuit mediated these interventions by reorienting subsistence activities in line with forms of commercial exchange operating beyond the scope of federal agencies.[40]

Inuit engagement with reindeer projects in Canada brings nuance to this point. Oral histories with Inuit herders in the Mackenzie Delta region conducted in the 1990s show frustration with the lack of individual freedoms within the herding lifestyle in the early twentieth century. Inuvialuit remember grueling periods of constant travel and wrangling, with few breaks for rest or community gatherings. These same Inuvialuit pointed to fur trapping as offering more fulfillment and dignity, even though it required similar commitments of labor and time. One conclusion to draw is that the hierarchy of supervision within herding—embodied by the foreman, the assistant supervisor, and the Sami instructors—irritated Inuit in the Delta. The private fur trade thus remained an escape from state-sponsored colonialism, even as it presented its own vulnerabilities. For Inuvialuit looking back, the risks of fluctuating market prices, competition with other trappers, and mercurial dealings with post managers trumped the rewards of working with reindeer—like secure housing, rations, and income. Not all Inuit from the Mackenzie Delta shunned herding, but those who did embrace it occupied a distinct class within northern society: they were known in the region as unsuccessful independent fur traders.[41]

Interestingly, scientists recognized these relations among the fur economy, Inuit resilience, and northern administration. In a 1935 meeting of the Interdepartmental Reindeer Committee, biologist Rudolph Anderson and anthropologist Diamond Jenness contrasted the functions of national parks with those of the Reindeer Grazing Reserve. The scientists agreed that national parks were designed to protect all wildlife in perpetuity. But in the reserve, herders had to hunt and trap, as they required fur for winter clothing that reindeer alone could not provide. Moreover, committee members decided to issue a hunting or trapping permit to any Inuit who wanted one, as "familiarizing the Eskimos with reindeer and the mere fact of their being on the Reserve" was bound to assist in their "education with regard to handling reindeer." In making these distinctions and concessions, Jenness and Anderson construed the western Arctic as a kind of liminal zone where bureaucrats struggled to operationalize authority. Even with protection ordinances, disciplinarian-supervisors, and Sami instructors, the willing participation of Inuit determined the success of the herding experiment.[42]

Thus when global financial contraction came to the Canadian Arctic, it also reconfigured the fate of reindeer herding and Inuit resilience there. In 1936, as bureaucrats in Washington contemplated the reorganization of reindeer herding in Alaska, fur trapper C. T. Pedersen sold his Canadian Arctic fur business to the Hudson's Bay Company. Like the Lomen Company, he found his integrated export operations proving less viable each year since 1929. The Hudson's Bay Company decided not to pursue trading operations in the Beaufort Sea, as had been established by Pedersen and his team of schooner-based traders. Rather than continue service between the Pacific coast and the Arctic coast, the Company relied on the Mackenzie River to connect the Arctic to markets and distribution centers in the south. These geographic and economic changes meant that all Inuit living along the north slope of Alaska and Canada west of Tuktoyaktuk who had relied on Pedersen for essential goods and equipment now had to find alternative means of procuring food, clothing, and cash.[43]

The removal of Pedersen's trading system from the Arctic led to broad changes in the region's settlement geography and the place of reindeer in Inuit well-being. With coastal trade pulled out from under them, residents in Barrow, Alaska, faced a food and fuel shortage. Many Native families moved out of town and into the interior, to use "oil soaked peat soil" as a fuel source and to rely on abundant ptarmigan, rabbits, fish, and reindeer for food. The remaining Barrow residents attempted to subsist off of the twenty-one thousand reindeer that survived on the range in the community's hinterland. Residents of nearby coastal towns at the mouth of the Colville River and at

Barter Island pleaded with Barrow residents for a share of these reindeer, and 3,500 animals were cut from the herd for use in these towns. By 1938, Inuit in Colville and Barter Island had killed all of these animals for food.[44]

In addition, many Inuit on Alaska's Arctic slope migrated into the Mackenzie Delta over the late 1930s and early 1940s to search for more suitable trapping conditions, continuing a trend in westward movements that began in the late 1800s with commercial whaling and disease epidemics. Trapping prices in the Mackenzie Delta remained high in this period, and Alaskan Inuit were willing to gamble with the dangers of relocating to Aklavik (a multiweek trip across thousands of miles of tundra) to continue a trapping lifestyle. The influx of Alaskan Inuit into the region concerned Canadian Inuit, however, who begged the Canadian government to establish trapping sanctuaries that prohibited non-Native residents—including Alaskan Inuit—from harvesting local game. The government did hear these suggestions but did not enact the proposed measures. Perhaps those in Ottawa were wounded by the fact that Inuit—a group many scientists and bureaucrats considered crucial to northern development—attempted to steer the process of natural resource management themselves, rather than take directions from southerners.[45]

Over the 1940s, scientists and bureaucrats altered their representations of reindeer herding in line with the changes in the fur economy and Inuit livelihoods. Biologists, who had in the mid-1920s yoked Canadian claims to the Arctic to the productivity and management of fur-bearing animals in the Mackenzie Delta, changed their tune. Now, herding and harvesting reindeer appeared as more stable than animal life cycles and the global fur trade. Regulating hunting of native species would not necessarily address the unpredictability of markets and nature, but building up reindeer as a subsistence base might. Such a "Native-run industry" gradually replaced the visions of a grand northern meat industry in governmental publicity of reindeer by midcentury. Behind the scenes, administrators increasingly targeted districts outside the Mackenzie Delta, especially those surrounding the Anderson and Coppermine Rivers, where Inuit had less experience with independent fur trading and were "more easily trained the natives from the Delta." This was code for Inuit who were deferent to white men, not necessarily those who performed duties according to any criteria of reliability. Indeed, foremen over the 1940s partnered with Royal Canadian Mounted Police in the central Arctic to act as human resource officers, selecting the "most reliable natives" available. They also instituted a "summer school" at Reindeer Station to expose Inuit youth enrolled in residential schools to the thrill and possibility of life as a reindeer herder. The majority of the Inuit hired on as herders in Canada thus came from towns like Coppermine and Cambridge Bay to the east of the Delta, and had previous experience working at police detachments. Western Arctic Inuit remained out of the grasp of southern bureaucrats, even as the reindeer experiment continued.[46]

And then, in 1944, a truly wild thing happened. On September 9, the "worst storm in years" blew over the Beaufort Sea coast, sweeping up the schooner *Cally*. On board were Peter Kaglik and Charlie Rufus, the first Inuit to graduate from apprentices to manage a herd, as well their families. Stanley Mason, the acting foreman of Reindeer Station, also died in the disaster. In a single day, nearly ten years' worth of investments—on behalf of scientists, Sami, Inuit, and the Canadian government—washed away. The loss of life on the *Cally* effectively reset the Canadian Reindeer Project. The process of hiring supervisors, recruiting Native northerners, and training them to be herders began anew. After 1944, only three other herds were spun off to Native hands, but each of these collapsed by 1956.[47]

DISTANT MEMORIES, ENDURING CONNECTIONS

The outcomes of these last herds point to still other forces, which pulled Inuit away from both trapping and reindeer and that signal another chapter in a century of outside intervention. Following World War II, and inspired by nascent Cold War geopolitics, bureaucrats and scientists framed the western Arctic as a strategic border zone with the Soviet Union and placed priority on defense initiatives. These choices relegated the notion of a wild Arctic, one in need of reindeer herding, as unimportant. Transformations proceeded accordingly.

Kittigaaryuit, where herders corralled reindeer in the Mackenzie Delta during summer, became home to a meteorological station for aerial navigation in the mid-1940s. In 1947, when Alf Erling Porsild visited the herd for a final time as a consultant for the commissioner of the Northwest Territories, the situation with recruiting Inuit apprentices had grown desperate. As an incentive for local participation, foremen lobbied departmental officers in Ottawa to give Delta Inuit more stake in decisions about reindeer. By the mid-1950s, though, hopelessness made this cooperative spirit a fugitive of history. As power brokers approved the construction of a string of radar stations across the Beaufort Sea coast and planned a new governmental center in the Delta, the feasibility of maintaining the reindeer program was brought into question. Some of the longest-tenured apprentice herders left Reindeer Station for jobs associated with these new developments, which paid more and expected less travel. In 1959, the government transferred the herd to private developers, maintaining a staff person at Reindeer Station for oversight. Today, between three thousand and four thousand reindeer continue to roam the tundra beyond the station. They remain complicated symbols of the Arctic's history of science and colonialism. The animals are managed by a Sami herder and owned by Canadian Reindeer, a company run by Inuvik resident Lloyd Binder. Binder's father, Otto, was an Inuk from the central Arctic who apprenticed in the Reindeer Project. Lloyd Binder's mother, Ellen Pulk, was a daughter of one of the original Sami herders contracted by Porsild in the early 1930s.[48]

In its rise, reindeer herding redefined the nature of Arctic colonialism. Naturalists, explorers, geographers, geologists, topographers, biologists, and anthropologists had been instrumental in surveying the Arctic for new lands, resources, and "lost" cultures before the Great War. But none of these specialists appeared as "qualified men" for the duties necessary in starting a government herd. In turn, when North American political leaders entrusted botanists with the Arctic's development, their ecological studies and concepts highlighted a set of unique issues in colonialism. Lichens and "reindeer mosses" were known to Canadian bureaucrats before ecological research, but *Cladonia*, "winter forage," "carrying capacity," and "tundra types" had not yet been quantified or made intelligible. The creation of a scientific grazing manager also reordered the positions of Inuit, herders, businessmen, researchers, and federal administrators relative to one another. Armed with charts and maps, the scientist-manager abstracted himself from the field operations of colonialist schemes, even as he governed them. The novelty of this science and its applications might be why some Inuit considered ecologists not practical reindeer men, but men with briefcases, issuing figures pulled from thin air.[49]

By midcentury, though, governmental reindeer schemes had lost their supporting scientific logic. This does not so much negate the historical significance of the herding experiment as underline it: relations among scientific knowledge, northern environments, and colonial power are both deep and delicate, lasting and yet historically contingent. Over the 1950s, US and Canadian officials, Interdepartmental Reindeer Committee members, and university researchers from both countries began wondering why this project—so destined to succeed—never lived up to its potential. Analysts hung the troubles of the project on Inuit culture, immature science, and poor planning. These autopsies of reindeer helped scientists and bureaucrats rationalize continued interference in the north, and instilled confidence that the Arctic could indeed be controlled. They also dislocated the past from the present. For governmental officials in the 1950s, experiments in taming the tundra were distant memories in an era of commanding a strategic terrain. For us, these same experiences offer enduring connections across a century of colonialism.[50]

CHAPTER FOUR

Strategic: Defense and Development in Permafrost Territory

In July 1947—after five years of trials and errors—Stephen J. Eszenyi became convinced that the most powerful military on the planet could not overcome the Arctic. As the officer in charge of the US Naval Arctic Test Station at Point Barrow, Alaska, Eszenyi had witnessed firsthand how the tundra thwarted the production of oil, the lifeblood of modern warfare. Bulldozers, tractors, and Weasels—a tank-like vehicle built during World War II for use in snow—could muster a top speed of only 5 miles an hour. Compensating for the rough terrain meant driving in second gear, which burned out transmissions. The actual drilling presented a whole other set of issues. At each location, engineers had to thaw the frozen surface before driving supporting piles, requiring expensive and heavy steam jet heaters. Once the soil warmed enough to allow excavation, the biting cold emerged from below, encasing the pile in rock-strewn ice. Exasperated, Eszenyi boiled down the situation for his navy superiors: "Pileloading in the Arctic is not the same problem as it is in the States."[1]

Failure was not an option. Alaska's north slope was a critical source of fuel for military operations in a northern theater of conflict, first during World War II and then in the Cold War with the Soviet Union. To ensure continental security, Eszenyi, along with the support of the US and Canadian governments, embarked on a radical course of action. He recommended that military forces ditch customary engineering principles and revamp their entire approach to defense. Rather than "combat nature and her forces by strengthening material," they should "cooperate with nature." Eszenyi proposed an extensive research program that investigated permafrost, the term coined for the foreboding subsurface conditions he found on the Arctic slope. He called for scientific investigations to determine methods for constructing drainage and sewage systems, roads, airfields, buildings and insulation, fuel storage pits, feed storage lockers, and, finally, piles. He was confident that results from such research would "make it possible for future economical planning."[2]

Eszenyi's proposal came to fruition. In 1947, the year he sent his memo, the Office of Naval Research created the Naval Arctic Research Laboratory, the United States' first laboratory in the Arctic. There, physical scientists from universities, governmental agencies, and armed forces in the United States and Canada worked side by side to figure out the dynamics of permafrost and how to work around them. In 1952, Canada's Division of Building Research and National Research Council opened a similar laboratory in the northern Northwest Territories. Like the US lab, the Northern Research Station at Norman Wells sat beside an oil production facility, this one operated by Imperial Oil, a subsidiary of Standard Oil. The more researchers and officials learned about the frozen soil, the more they saw knowledge about it as foundational to both the Arctic's defense and its long-term development. In addition to its immediate military applications, permafrost science promised to unlock Arctic oil and natural gas deposits. Boosters hoped these northern resources could meet skyrocketing consumer demand for petroleum products in metropolitan North America. As territorial officials in Alaska curried favor with private investors, the government of Canada sought a suitable site for an administrative hub for a future Arctic oil economy. In 1959, the same year Alaska earned statehood, Canada completed Inuvik in the heart of the Mackenzie Delta. Both the US and Canadian governments seized this moment, and the fund of scientific knowledge about what lay beneath the tundra, to lease millions of acres of the western Arctic to oil companies. Between 1944 and 1959, scientific understanding of permafrost catalyzed another round of transformations in the far north.

That scientific representations of a strategic Arctic might inspire such interventions was nothing new to the region by midcentury. But this chapter of colonialism's northern history does stand apart from its predecessors for two significant reasons. First, the wartime context and innovations in transportation technology made it possible for governments to stage operations in the Arctic that were far more complex and widespread than in previous years. Meanwhile, as scientists, militaries, and oil companies moved north, Inuit had fewer opportunities for mobility and adaptation. The means of production developed by outsiders—whether of science or oil—excluded Native northerners like never before. Field-workers and engineers flew in by airplane to towns and laboratories they built to extract resources that Inuit did not ordinarily use. Yet even if Inuit did not imagine themselves within the world being created by southerners, they could hardly avoid participating in it. While the United States and Canada established an Arctic oil economy, the world Inuit had built deteriorated. In the 1950s, fur-bearing creatures became harder to find, markets for fur evaporated, and the Hudson's Bay Company converted its Arctic fur posts from fur trade centers to retail outlets. For both outsiders and Inuit, the 1950s were a turning point, when the machines and methods of colonialism became the vehicles of cultural survival.

"A JUNKYARD MONUMENT TO MILITARY STUPIDITY"

The attack on Pearl Harbor and the occupation of Alaska's Aleutian chain by Japanese forces threw the geopolitical importance of the north in sharp relief for North American bureaucrats. Both the US and Canadian governments had recently completed the Northwest Staging Route, a string of airfields connecting Edmonton with Fairbanks, Alaska to supply lend-lease aircraft to Soviet forces on the eastern front. Now, with immediate threats just offshore, these airfields became a central piece of a larger system of infrastructure for the corner of the continent. In June 1942, President Roosevelt approved the US Army's proposal to construct a transportation corridor across the subarctic.[3]

Over the next twenty months, US and Canadian governments and private enterprise built transportation facilities on the frozen earth, foreshadowing the kinds of activities that would trouble naval officers like Eszenyi five years later. The US Army Corps of Engineers, together with the joint venture Bechtel-Price-Callahan and Standard Oil, cleared a 1,700-mile roadway to supply logistical support and ready access to northern airstrips. They also assembled a 1,000-mile pipeline from Norman Wells, in the interior of the Northwest Territories, to Whitehorse, Yukon Territory—a protected source of oil to fuel defense initiatives in Alaska. In 1943, as vehicles drove across the Alaskan Highway for the first time and oil began flowing through the CANOL pipeline, more than 450 aircraft flew overhead every month on their way to the far northwest. Meanwhile, the forty thousand soldier-workers and engineers called on to complete these projects returned south, or headed off to other outposts across the world. This megadevelopment—large in its size, extent, and capacity—was a sign of things to come in the Arctic over the 1950s.[4]

The long-term maintenance of the CANOL pipeline proved elusive. The rush job encouraged shoddy craftsmanship and riddled the pipeline with countless needs for repair. Thirteen months after it was completed, the United States and Canada closed CANOL, reopening the tactical wound of reliable, invulnerable oil. Observers in the civilian sector rubbed salt in it. When news broke that the mission cost more than five times its twenty-million-dollar estimate, the CANOL project became known as a "junkyard monument to military stupidity." In this way, military activities in the subarctic during World War II had lasting ideological legacies, to complement their material effects. CANOL was a symbolic failure to subdue nature in the name of peace and progress.[5]

Against this track record, and with the same imperatives for defense, the United States took bolder action farther north. In 1943, the Department of the Interior issued Public Land Order 82, which withdrew all of Alaska north of the Brooks Range—about 49 million acres—for "the prosecution of war." In the same year, the secretary of the navy and the US Bureau of Mines sent a reconnaissance team to the abandoned Naval Petroleum Reserve no. 4 along the western Arctic slope of Alaska. Created in 1923, "Pet 4" had been imagined as a ready source of crude oil in times of short supply or emergency. While geologists in the 1920s had concluded the petroleum deposits were remote and not extensive enough for commercial exploitation, the vagaries of war raised the value of the land. By the next year, the navy arrived on site along with Arctic Contractors, a civilian group hired to help extract the oil and natural gas.[6]

NECESSITIES OF A WAR IN THE COLD

The US Navy and Arctic Contractors spent the next decade scouring Pet 4 for oil, but not to support combat in Europe. While the use of nuclear warfare brought World War II to a swift end for the United States, it also brought about another international conflict with a circumpolar neighbor. With the Soviet Union rising as a rival, and the advent of airplane and missile technologies, the far north appeared to North American strategists as a thin barrier between two superpowers. The long-standing confidence in Alaska's "vastness" as its own defense had been shattered. If combat were to break out in the Arctic, how would armed forces respond? An airlift-style operation that had proved successful during World War II was too expensive. Troops would have to learn how to fight on the ground. The air force, navy, and army would need to develop transportation and survival capabilities in cold temperatures, darkness, and stretches of land without commercial services.[7]

US governmental and military personnel thus performed massive construction projects to aid Cold War defense efforts. Pet 4 was the first of these. By 1953, after ten years of operation, the navy could report it had covered a staggering amount of ground. They had taken aerial photography over 112,000 square miles of the reserve, conducted initial tests of oil productivity in over 60,000 square miles of land, and mapped another 60,000 square miles of the reserve's geological foundations. They drilled more than seventy test wells. These efforts led to the discovery of three active natural gas wells, which supported all the defense-related operations on the north slope of Alaska. These results cost the navy more than $50 million. The scale and intricacy of these activities far outpaced previous developments and replaced the Canadian Arctic Expedition as the largest, most expensive research effort in northern history.[8]

Yet, as Stephen Eszenyi pointed out, the success of the navy in maneuvering the Arctic—and thus protecting the continent—depended not on financial outlays, but accounting for the environment. Arctic surface and subsurface conditions complicated transportation, construction, and oil extraction in several ways. First, the north slope of Alaska and Canada had low total annual precipitation, which forced operators to attend to water resources. Point Barrow, the site of the most elaborate facilities, abutted the ocean and could take advantage of marine shipping. The navy placed the main subsidiary camp on the Colville River for similar reasons. Still, the tundra offered no available groundwater, and transportation of water at any distance risked freezing. The installation and maintenance of sewers was costly and difficult because pipes froze if not heated. Yet if heated, they altered the grade of the land by melting the frozen subsurface material.[9]

Second, the seasonal differences between summer and winter frustrated any kind of oil exploration activities and, again, put emphasis on the need to understand the physical environment. Winter was long, with few frost-free days. This created a long heating season and unusual demands on building operation, design, and costs. The navy initially created foundations by pouring concrete slabs on gravel pads. But within a month, the slabs began "to crack and settle unequally" because they had melted the ice beneath the soil. When officials believed the "equilibrium had been reached" they repoured the floors, but soon experienced the same problem. When winter was over, summer created other issues. Navy operators initially began using Weasels and LVT's (landing vehicles tracked), but found that these would routinely get stuck in the "wet, soggy tundra." They learned that transportation was best suited for winter, when the land was frozen solid and could bear the weight of machines and personnel. Between December and May in 1953—or, after freeze-up in the fall and before the thaw in late spring—Pet 4 operators logged more than 2.5 million ton-miles of overland freighting.[10]

Third, and finally, the conventional communication and material technologies employed to extract oil were out of sync with the Arctic's geographical and physical constraints. Consider the drilling of a 5,000-foot test hole 200 miles from Point Barrow, the main camp. This was a routine activity during the ten-year lifespan of Pet 4, but by no means uncomplicated. To begin, a decision had to be made to drill the hole at a location somewhere in the 70,000 square miles of "Arctic wilderness" that composed Pet 4. This decision had to pass through an operating committee, itself a maze. But because little geological or physiographic information was initially available on the land, preliminary surveys had to be completed before the drilling could commence. Once sufficient data was in hand, the operating committee would debate a proposal in November—after the summer field season was over. Then, equipment and supplies would have to be listed and procured during winter and moved to a shipping point on the Pacific coast—either Point Hueneme, California, or Seattle, Washington—to arrive on the beach at Point Barrow by the following August. This material needed to be organized, perhaps winterized, and prepared for shipment before being moved by caterpillar train in February. Upon delivery, it would remain at the drill site until April when it could be assembled and put to use. A year and a half had elapsed between the decision to drill and the commencement of drilling.[11]

Facing the possibility of another CANOL pipeline blunder, the navy sought much more scientific support to deal with the engineering problems presented by Arctic nature. Like Eszenyi, military officials and governmental leaders equated knowledge about permafrost with defense strategy because it formed the basis of all human activity in the north. According to one report, "there is probably no other single favor in the environment that will cause so much frustration, require so many established practices to be abandoned, new techniques to be learned, and be at the root of so many practical problems." In early 1947, the US Navy established the Naval Arctic Research Laboratory, the first scientific station in the North American Arctic (figure 11). The first structure was a navy-built 20 x 40 foot Quonset hut refashioned as a laboratory, with temperature control chambers. By 1968, the Naval Arctic Research Laboratory consisted of forty-one laboratory facilities, with sleeping space for more than eighty people on a footprint of more than 45,000

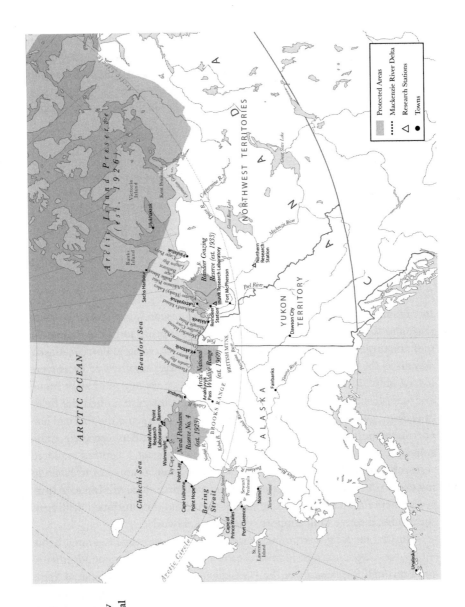

FIGURE 11. The west-
ern Arctic in 1959. The
Naval Arctic Research
Laboratory was estab-
lished next to the com-
munity at Point Barrow
and just above the Naval
Petroleum Reserve
no. 4. Map by Morgan
Jarocki.

square feet. As a testament to the necessities of permafrost knowledge for a war in the cold, the Naval Arctic Research Laboratory's budget over the 1940s and 1950s was equal to all other Office of Naval Research projects carried out in the subarctic and Arctic combined.[12]

THE THERMAL BALANCE OF POWER AT THE NAVAL ARCTIC RESEARCH LABORATORY

The temperature was in the 50's and all around was the noise and bustle of an oil-exploration camp. Caterpillar tractors churned the soft sand as they hauled equipment to storage areas. Weasels (M29C), those small tracked vehicles so useful in the Arctic, seemed to be scooting in all directions on a variety of missions. The landscape was dotted with 56-gallon fuel drums, that ubiquitous trade mark of the American developer in out-of-the-way places all over the world.

JOHN C. REED AND ANDREAS G. RONHOVDE[13]

In commemorating the birth of US laboratory research in the Arctic, scientists John Reed and Andreas Ronhovde linked the project of studying nature with the project of drilling for oil. For researchers arriving in Barrow, Alaska, as well as for many other observers at midcentury, these two projects were inseparable. While the Naval Arctic Research Laboratory pursued a broad and interdisciplinary research program, singular interests united the fields of ecology, geology, and geophysics. These scholars targeted the thermal balance in the permafrost, or the exchange of temperatures between surface and subsurface that govern the stability of ground material. Scientists explored research questions and methods on thermal balance as a means of facilitating extraction of petroleum resources from the north slope of Alaska. In turn, researchers found the logistics of the oil camp, and the continued drilling of the tundra, a convenient support system for fieldwork.

Consider the longest-running program at the Naval Arctic Research Laboratory, the "Arctic Ice and Permafrost Project." This project came together in 1949 through the support of the US Geological Survey, the Bureau of Yards and Docks, and the Office of Naval Research—all agencies with significant interest and influence in Pet 4 operations. The Arctic Ice and Permafrost Project maintained a variety of topical investigations, from creating detailed maps of the distribution of permafrost in northern Alaska to the study of effects of "continuous" permafrost on human geography in northern regions (as contrasted with "discontinuous" areas with only sporadic permafrost). The staff of the project also compiled a bibliography of publications relating to permafrost to supplement the work of the US Army, Air Force, and the US Geological Survey. But the most significant lines of investigation—those that gathered the most funding, the biggest investments of time and labor, and the most interest from governmental authorities—were examinations of the "economic aspects" of permafrost. These entailed the "seasonal freezing and thawing as applied to living conditions, house building, water supply, underground cold storage, sanitation, roads," as well as airfields, telephone and power lines, and drilling in northern Alaska.[14]

The fieldwork of scientists in the Alaska Ice-Permafrost program shows the mutual relationship among the accumulation of knowledge about heat transfer and the elaboration of the military-industrial complex in the Arctic. Geologist R. G. MacCarthy, a professor at University of North Carolina–Chapel Hill, addressed the temperature profile and moisture content of permafrost by working closely with the navy operators and the oil wells they punched throughout Pet 4. Between February 1949 and July 1959, MacCarthy sunk more than thirty thermistors—cables that offered indirect measurements of temperature through measurements of resistance—into test wells throughout northeast Alaska. These wells varied in depth between 200 and 1,000 feet, offering MacCarthy access to the subsurface world. By creating a database of recordings at various depths, the geologist hoped to determine the character and extent of the "active layer," the area between the surface and the solidly frozen earth beneath that was subject to changes in temperature. MacCarthy also used resistance to measure soil content. He placed four electrodes into the earth and passed a known electrical current through two of them. In this way, he could measure the change in potential between the remaining two electrodes—which became an indirect measurement of soil moisture. In turn, he used soil moisture to indicate the kind of materials in the ground, whether silt, sand, or gravel.[15]

Such work would have been possible without the navy's infrastructure in camp facilities and regular shipping schedules from Point Barrow throughout the Naval Petroleum Reserve—but much more difficult. One of MacCarthy's colleagues, Robert F. Black, described his work on ground ice as "labor in the wilderness" and complained about the time investments of doing field research on permafrost. Any fieldwork took "two to four times longer" than in the States and involved incredible effort. "How many places do you find it necessary to chop frozen ground and ice for as much as 4 hours to get a decent sample for study?" he asked. MacCarthy and others could circumvent these issues by following the trails blazed by the navy and their private contractors in boring holes in Pet 4 (figure 12).[16]

FIGURE 12. US Navy contractors move a drill rig between test wells outside of Barrow in 1950, paving the way for geological research on the subsurface environment. Credit: US Geological Survey.

In turn, research with thermal resistors helped mediate concerns from navy oil operators about oil production. Arctic Contractors, one of the firms working with the navy, had reported that the casings of its wells were mysteriously collapsing. In December 1950, the director of Naval Petroleum Reserve called a special meeting "on the subject of production in and through the permafrost." Participants noted it seemed either an outside force from the ground supplied a crushing pressure, or that the pipes were put under stress because of the freezing temperatures. Temperature studies of wells became one of only three year-round projects ever conducted at the Naval Arctic Research Laboratory, and continued for a dozen years. According to directors of the laboratory, the insights from this research "proved critical" to the navy's success with natural gas production in the 1950s.[17]

As they built up a database of observations, scientists of the Arctic Ice and Permafrost Project gathered a more comprehensive view of the north.

Scientists applied measurements of soil moisture to map the distribution of permafrost throughout the territory of Alaska, since its phenomena occurred beyond the Arctic. If the permafrost consisted of solid or fragmented rock, or of dry and well-drained sand and gravel, the subsurface dynamics were scientifically interesting but of no unusual engineering significance. However, if the permafrost consisted of waterlogged silt—and thus had a higher value in resistivity studies—the ground might create a potential nightmare for construction and travel. In these cases, the thermal balance of the frozen soil and water could be easily disrupted, and previously solid permafrost would assume "the consistency of soup when thawed." Through soil moisture, permafrost scientists could help better locate power plants, barracks, hangars, and even hospitals and avoid blunders like the CANOL pipeline.[18]

The application of permafrost knowledge attracted an audience beyond the US military and the confines of Pet 4. In 1954, John C. Reed, the geologist who eventually wrote the history of the Naval Arctic Research Laboratory, penned a short memo titled "Pet 4—A Key to Arctic Operations" to the Office of Research and Development in the Department of Defense. This document shared the "rich store of information" gleaned from research in the Arctic in hopes of systematizing it for "maximum advantage" by other operations and corporations. Reed distilled his recommendations into a short list of "Do's and Don'ts" for potential future operators in permafrost country, all of which emphasized the importance of environmental features. Interested parties should "make maximum use of favorable seasons," "provide space for air circulation under buildings" to prevent the melting of permafrost, and avoid grading roads or runways "without removing sufficient frozen silt." Military officials applied Reed's list in future polar megadevelopments, including the Distant Early Warning Line—a string of radar stations built across Arctic North America between 1953 and 1958—as well as in related facilities and operations in the Antarctic. When the director of the Naval Arctic Research Laboratory sent an abstract of permafrost research results to colleagues in Canada, there was great interest from the Defence Research Board and the Department of Transport.[19]

TERRAIN INTELLIGENCE

Permafrost science did not have equal application in Canada. It had more, for the simple fact that the extent of the Dominion's northern territory trumped that of the United States. More immediately, though, regional particularities flummoxed Canadian officials and made permafrost science a foundational concern. Researchers found that the greatest concentration of the waterlogged permafrost—that troublesome for construction—existed along the coast of the Beaufort Sea and in the Mackenzie Delta in the western Arctic, the same area underlain by oil and natural gas. And since the north slope of the Yukon Territory and Northwest Territories fell in the same geological province as Pet 4, it seemed probable that these areas contained similar amounts of hydrocarbons as the navy found in northwest Alaska—an estimated 100 million barrels of oil and 900 billion cubic feet of natural gas. Indeed, bureaucrats within Canada's northern administration estimated the sedimentary basins of the Yukon and northwestern Northwest Territories to house a reserve of oil and gas that was seven times the discovered materials in western Canada. Not surprisingly, Canadian governmental and military leaders followed the Arctic Ice and Permafrost Project closely and sought to replicate its efforts in their own north. In the early 1950s, the National Research Council of Canada's Division of Building Research put full energies behind what it called "a new field of study—permafrost."[20]

Canada's foray into permafrost science came with the Canadian Arctic Permafrost Expedition of 1951, a joint venture between Purdue University, the US Army Corps of Engineers, and the National Research Council of Canada. The goal of the survey was to perfect new methods for Arctic engineering. Similar to the ways scientists used soil moisture as an index and map of permafrost, expedition researchers wanted to develop techniques for determining soil types and subsurface conditions by combining aerial photography and soil sampling. Such work had been carried out in Alaska over the end of the 1940s along the Alaskan Highway, but permafrost scientists sought additional field sites to improve their approach. The route of the Canadian Arctic Permafrost Expedition followed major arteries of transportation in the Mackenzie River valley, not only to facilitate the logistics of data collection but also to understand the relationship between terrain features and infrastructure so as to foster more efficient construction. In the movements of the Canadian Arctic Permafrost Expedition, then, the interests of science and colonialism and the United States and Canada converged on one another.[21]

The itinerary for the expedition underscores the particular material and ideological relations of science in the Arctic after World War II. In Western Arctic Expeditions during the early 1900s, scientists relied on schooners, whaling outposts, dog-sledding, and trading on the Beaufort Sea coast to collect observations and specimens. The Canadian Arctic Permafrost Expedition of 1951 ran in an entirely different fashion. Some members of the group, including ecologist Alton A. Lindsey, flew from Chicago to Edmonton, and then Edmonton to Yellowknife, which had become a gold-mining center on the banks of the Great Slave Lake in the southern Northwest Territories over the 1930s. Others, like professor of engineering at Purdue, K. B. Woods, and his colleague James Shepard, drove from Saint Paul, Minnesota, through to Edmonton and beyond, using the Alaskan Highway at times, in order to view the landforms of the Canadian west for comparison with the north. Once in the Mackenzie Delta, the rhythms of summer fieldwork seemed more consistent between the early and mid-1900s. Shepard cruised aboard a steamer, the *Cheechako*, from Fort Good Hope to Aklavik. His hopes of studying the vegetation communities along the banks were dashed as the weather made it "hard to do a lot of good work." Similarly, Alton Lindsey harkened the work of reindeer biologists when he completed some ecological surveying by canoe in the vicinity of Reindeer Station (figure 11).[22]

The major difference between Western Arctic Expeditions in the first two decades of the 1900s and the Canadian Arctic Permafrost Expedition of 1951 was not that scientists had access to different modes of transportation, though. It was that different ways of moving through nature offered different ways of knowing the environment. The decisions on driving, canoeing, cruising, and flying were made intentionally to create a picture of the total Arctic environment, to combine the perspective of the landscape as seen from a seat of an airplane with the detailed picture of the patch of lichens, as seen on the banks of the Mackenzie River. The view from the plane allowed for the identification of what one scientist called the "patterned ground," or the different "circles, nets, polygons, steps, and stripes" unique to permafrost territory. The view from the canoe allowed the correlation of each of these patterns with a set of plant communities and soil samples, and possibly, an underlying ecological or geological process that dictated the pattern.[23]

Scientists continued to combine and refine these perspectives through the completion of the Canadian Arctic Permafrost Expedition. They compared the six thousand photos they snapped during fieldwork with the available database from Purdue University, the US Army Corps, and the Royal Canadian Air Force. These comparisons provided confidence that "sufficient data" had been collected for application in developing the Canadian Arctic. The Canadian government responded by investing further still in permafrost research. In 1952, the Division of Building Research opened the Northern Research Station at Norman Wells, a town at the end of the line for wheeled aircraft in close proximity to the oil refinery first used to pump crude along the CANOL pipeline. John Pihlainen, the sole representative of Canadian science on the Permafrost Expedition, became the manager of the station. Like

his colleagues at the US Naval Arctic Laboratory, Pihlainen leveraged oil development as a research opportunity. When Imperial Oil built new facilities
at Norman Wells, he collected data on "site preparation"—like how to clear
vegetation from the surface and how to fill foundations with gravel. Borrow-
ing company equipment, he took frozen drill cores throughout the Mackenzie River district to understand the physical characteristics associated with
permafrost—such as climate, location, and composition. And, in his office, he
polished a scheme of air photo interpretation by marking up transparency
overlays to trace landforms, vegetative cover, borehole locations, and soil
test results. From these experiences, Pihlainen developed presentations and
publications for scientific audiences, which became standards in the field of
permafrost science.[24]

By the mid-1950s, these methods for studying permafrost had provided
administrators with the engineering principles naval officer Eszenyi desper-
ately sought in the 1940s. Engineers and scientists referred to the knowledge
they helped create as "terrain intelligence." Terrain intelligence could help
armed forces understand which areas of the ground would be impossible to
move vehicles. It could also pinpoint where to find gravel in order to build a
road or airfield in an emergency situation. Such capacity was crucial to any
kind of development. Through permafrost research, US and Canadian decision makers reclaimed some authority over an Arctic environment that had
foiled their plans on the ground during World War II. Engineers claimed a
terrain intelligence so detailed that, given a few photographs of an unknown
location, they could "name the area anywhere in North America."[25]

THE MAKING OF EAST THREE

In the late 1950s, these engineering principles would be put to work in one
of Canada's most ambitious plans of the twentieth century—building a governmental hub in the Mackenzie Delta. In 1953, the Advisory Committee on
Northern Development met to discuss the need for a new Delta community
to serve as the "educational, administrative, and welfare centre" of the Arctic
as well as the center for the petroleum industry. This meeting was prompted
by research performed by permafrost scientists in Aklavik, the existing economic and governmental nucleus in the region. Given aerial photographs
and soil sampling performed locally, scientists concluded that Aklavik could
not house industrial development. It rested on organic silt with "very high
moisture content," with 60 percent of the frozen fine-grained soil consist-
ing of ice. If the town were to be developed—through land clearing and the

installation of heated buildings—ground subsidence would ensure with "serious results." Moreover, the subsurface conditions would prevent the construction of a major airstrip and inhibit adequate draining for sewers or water mains. Even the Mackenzie River, which had given rise to the fur trade that made Aklavik the "fur capital of the world" and had for so long been integral to Canadian and British empire in the north, became a liability. Because the town was situated on a river bend, there was no "natural land for the town to expand to," which was necessary "to fulfill its role as the centre for administration for northwest Canada." The future of Canada lay elsewhere in the Arctic, in a town site that maximized the potential for oil and gas development while avoiding, as much as possible, the physical limits permafrost placed on development. If this is what naval officer Eszenyi meant by cooperation with nature in 1947, it looked a lot like colonialism by 1954.[26]

In response to this data, the Advisory Committee on Northern Development created a survey team to find possible sites in the Mackenzie Delta for a new town. This survey team featured three permafrost scientists in R. F. Legget of the National Research Council, R. J. Brown of the Division of Building Research, and John Pihlainen, of the Northern Research Station. While fieldwork commenced in 1954, scientists began their survey without leaving Ottawa. Looking at recent aerial photographic records and soil samples from the region, they identified twelve sites in the Delta for follow-up field investigations. Photographic keys remained instrumental in governmental plans at midcentury because they allowed scientists and bureaucrats to match particular landscapes with particular goals for "modernization." After their follow-up survey, permafrost scientists described the requirements for the new town, which they created in negotiation with the federal government. The location would have to be suitable "from economic and social points of view," meaning that it would be situated near the abundant oil and gas fields of the Mackenzie Delta and as close as possible to a navigable waterway. The new town should "allow installation of permanent sewer and water systems, building foundations, and roads" and have a "first class airfield" nearby. It should provide for convenient disposal of sewage and have suitable public water supplies. All of these principles corresponded with the problems posed by permafrost. Indeed, the authors commented that it "would be convenient if gravel and sand sources were nearby for construction purposes" given that hospitals, roads, and airfields could not be placed directly on the land surface.[27]

As the site selection process allowed scientists to apply terrain intelligence, it also became another venue through which to elaborate permafrost science. When considering the possible sites from the ground, scientists and engineers drew from knowledge gained about permafrost from the US military and civilian agencies. The field team replicated the combination of aerial reconnaissance and ground-truthing carried out on the Canadian Arctic Permafrost Expedition. Beginning in March allowed the team to observe potential town site areas "under late winter conditions, during spring break-up, and during the summer season." Using a helicopter to do an initial flyby of the twelve sites resulted in a paring down to six, because the others were "obviously unsuitable" when examined in the field. For the remaining sites, the team established a testing procedure that included topographic and hydrographic studies, as well as soil sampling and more aerial photography. They set up a base camp that could be transported by "tractor train, dog team, and by a small barge and scows," to each prospective town site as the season progressed. Its portability revealed that science and technology had adapted, in part, to the special problems of the Arctic's seasons and environment that had incapacitated US armed forces in earlier years.[28]

Soil samples from drilling became crucial in making a final decision. Samples were taken to the Norman Wells laboratory for routine engineering tests, which included moisture content and grain size. These tests showed that one site, "East Three," had a preferred diversity of soil types (for locating gravel and sand for construction purposes) and extent of glacial till, which "will yield more satisfactory foundation conditions than at the old site." East Three also had the benefit of topographical relief, which would ensure adequate drainage for all the proposed uses of the town. Moreover, the site was aesthetically pleasing. It was beautiful, permafrost scientists wrote, "especially in summer, and spreads over rolling hills that are enlivened with small lakes." The engineers could easily see with "careful town planning" that the future of East Three "should be as attractive physically as the old Aklavik was disappointing."[29]

The Advisory Committee on Northern Development wasted little time in forwarding the survey team's recommendation. After the minister of northern affairs visited the site—which marked one of the "first occasions when the Federal Minister responsible for Northern Canada has been able to study in the field"—the cabinet officially decided to relocate the town of Aklavik to East Three on November 18, 1954. The survey team considered its work a success—a demonstration of "what can be achieved by real team-work in the interests of the awakening North."[30]

"HUB OF THE WORLD?"

In 1961, when Canadian prime minister John Diefenbaker landed at the airstrip in Inuvik—the new name for "East Three"—the gravity of the moment was not lost on him. As the first prime minister to visit Canada's Arctic territories, the native of Saskatchewan made connections between his trip and that of another influential western Canadian, Sir John A. Macdonald. Under the midnight sun, Diefenbaker recalled the "famous journey" made by Macdonald to the shores of British Columbia via the newly completed transcontinental railway in the late 1800s. Both men, Diefenbaker said, sought to spark "the imagination of Canadians" with a "vision of a greater Canada." For the midcentury political leader, governmental-sponsored infrastructure projects in the Arctic—like Inuvik—would link the country's northern oil resources and its southern-based population to create a more expansive geographic and political territory, not unlike the confederation of Canada in Macdonald's time.[31]

In advocating these kinds of development in the Arctic, bureaucrats in the late 1950s conjured visions of the strategic Arctic that did not necessarily invoke its military importance. After World War II, a growing appetite for petroleum in North America and a difficulty in sustaining access to petroleum resources made the Arctic an appealing source of energy. Total demand for oil and natural gas in North America reached a record high in 1955, and again in 1956. Meanwhile, the Korean War (1950-53), the shutdown of Iranian oil in 1951, and the Suez Conflict in 1956-57 convinced political and economic elites that a steady stream of oil was no longer just a necessity of war, but was crucial for economic progress. As one governmental memo put it, the "problem of an adequate supply of petroleum" was far more than a military issue, because it involved "industry and transportation of the whole country." Given these conditions, domestic supplies in the Arctic appeared precious, useful, and even attainable, especially with the recent accumulation of knowledge about permafrost. In 1957, Atlantic Richfield struck oil on Alaska's Kenai Peninsula in the southeast, which sparked a flood of capital investment into the territory that echoed the Klondike gold rush of the 1890s. By the close of the decade, oil had supplanted coal and gold as the most important mineral product of the western Arctic.[32]

As Diefenbaker's trip to Inuvik suggests, hopes for capitalizing on the Arctic's abundant mineral reserves went hand in hand with desires for enhancing governmental capacity in the region. Bureaucrats pointed to the Arctic as an "underdeveloped" location, one that required more localized state agents to help lift the region out of poverty and take advantage of its natural wealth. "The Canadian north is one of the last great undeveloped regions on this globe," announced high-ranking northern administrator Gordon Robertson. "There are few other parts of the world of great size that have not been occupied—insofar as they are capable of it—and to a substantial degree exploited." Robertson was the right-hand man to Diefenbaker on the prime minister's Road to Resources program, which sought to establish transportation infrastructure in the north to facilitate the extraction of its natural resources.[33]

The discussion of transportation facilities at midcentury thus acknowledged the Arctic's similarities with other colonial locations—but only as a means to proclaim a faith in development led by the academic-military-industrial complex. Consider *The State of Alaska*, the book written by governor of the Alaskan territory, Ernest Gruening, as a tool to lobby for statehood in the 1950s. In a chapter titled "Transportation: Tangled Life Lines," he pinned the potential for statehood on the creation of shipping routes, airfields, highways, and railways. Gruening explained that a "lag in Alaska" was due to the "distantly man-made obstacles" of federal bureaucracy and that local investments in "those vital arteries through which the lifeblood of commerce would course" would guarantee a bright future for the state. Without such strategic construction, he said, the economy of Alaska would continue to resemble "the emerging nations of Latin America or Asia." Here, as in many other utterances by elected officials at the time, was the "old story of any frontier area." Transportation facilities were too expensive to build without a viable market for minerals, and resource exploitation by private enterprise would not proceed without transportation infrastructure in place. Both US and Canadian officials called upon the public to break this "vicious circle" by supporting governmental investment in Arctic roads, airfields, and towns—and thus permafrost science, too.[34]

For their part, private industry sat poised to enter the Arctic. In 1960, Imperial Oil issued a "Report on the North" meant to pique shareholder interests in investment. The authors noted that the year marked the end of a "decade of feverish activity" in setting up the foundations of an oil economy in the north, through airstrips, the Pet 4 facility in northwest Alaska, northern laboratories, and other administrative centers. For Imperial Oil, these signs suggested the start of a new world order centered on the North Pole. The opening essay, titled "Hub of the World?" featured a polar-projection map, reorienting the reader's view of the earth on the Arctic Ocean. The image and the text supported the idea that the far north was transforming into a lynchpin in the global oil economy. Similar ideas trickled through discussions in meetings of the Chambers of Commerce in Fairbanks and Anchorage at the time and the National Northern Development Conferences that began in Canada in the late 1950s.[35]

Ultimately, with governmental support, permafrost science helped realize these dreams of Arctic oil. In 1954, the US Department of Defense made all of the records and geophysical data relating to oil explorations at Pet 4 and the Naval Arctic Research Laboratory available for public inspection both in Washington, DC, and in Fairbanks. It also ordered the US Geological Survey to publish all information obtained by the navy and its researchers between 1944 and 1953 because "elements of the industry and others have shown interest in further exploratory work in Alaska." In 1958, the US government returned lands of Pet 4 to public domain, which caused "feverish speculation and exploration despite the remoteness of the area." By the end of the decade, more than twelve million acres of Arctic land had been leased by the state of Alaska to multinational oil corporations.[36]

Similar events transpired across the border. In 1958, at the First National Northern Development Conference in Edmonton, Alberta, Canadian power brokers proclaimed the Arctic as the scene of the "world's greatest search for oil and gas." Exploration had covered more than seventy million acres in the Canadian Arctic since 1948, and the extent of leases in the Beaufort Sea coast region had doubled since 1950. The first resolution of the conference was to request the Canadian government share more information about permafrost with private agencies, as had been done in Alaska four years before. In response, after completing construction at Inuvik, the Division of Building Research shut down the Northern Research Station and moved it to the new Arctic town, opening the Inuvik Research Laboratory in the early 1960s as a coordination center for oil exploration and development.[37]

In his dedication speech for the community in 1961, which he delivered across the street from the as-yet unfinished lab, Prime Minister Diefenbaker encapsulated the moment. As a town built by the government, because of permafrost knowledge, for the express purposes of oil production, Inuvik was a place "with no past to leave behind—only a future to look forward to."[38]

IN THE ARCTIC, ON THE OUTSIDE

He was wrong, of course. Outside interventions in the Arctic did have a history, one that was more obvious to Native northerners than southern bureaucrats. When we consider what Mackenzie Delta residents thought of the activities in the region, we see that the history of Arctic colonialism hit a turning point in the 1950s—but not only because of the size and complexity of defense and development. Both scientific research and governmental capacity-building at midcentury marginalized Inuit in ways more direct and more infuriating than in the past. When combined with contemporary changes in the ecology and economy of the fur trade, Inuit found their ability to mediate colonial forces severely compromised.

The airplane and helicopter strained relations among researchers and northerners. These technologies relieved field-workers from establishing extensive and regular relationships with locals as guides, interpreters, and informants. Permafrost scientists in particular could produce knowledge about the Arctic environment without Inuit expertise and apply that research in governmental construction projects without consulting locals. Inuit took notice of these changes. In a 1959 poem titled "The Northern Plague," one Native northerner likened governmental scientists to mosquitoes. They arrived in summer "in lusty swarm" and were just as annoying, "flitting about in copters eerie." The poet also noted how scientists could collect natural resources—whether fish, game, or minerals—for the purposes of knowledge production and modernization. Yet the very government that sponsored such science prohibited local residents from these same activities. "All this in the guise of science," went the poem, "but we natives require a license."[39]

Even when scientists consulted directly with Native northerners about their plans, negotiations were one-sided or, worse, two-faced. The permafrost scientists with the Aklavik relocation survey team established a "local advisory committee" to help select the final site for the Canadian government's new Arctic hub. This body had disproportional representation from non-Natives, though. On the committee sat two missionaries, a Royal Canadian Mounted Police officer, a well-regarded independent fur trader, a government agent stationed at Aklavik and Reindeer Station, and two Aboriginal men, Charles Smith and Frank Carmichael. Beyond the non-Native majority, however, sat a more fundamental problem with the committee: it was never meant to incorporate indigenous voices. According to anthropologist Peggy Brizinski, who interviewed Mackenzie Delta residents in the early 1980s about the creation of Inuvik, the local advisory council was designed to "channel information and persuade the people to move." Native northerners remained outraged by this facade of cooperation for decades to come.[40]

Without a say in whether or not Inuvik would be built, or where it would be located, Native northerners turned to the decision of moving to the new town. This presented a series of devil's choices. As "The Northern Plague" poem suggests, living in Inuvik invited closer governmental management of

Native livelihoods. Given the town's location within the Reindeer Reserve, Inuit would be required to acquire a permit to hunt and trap. Many Aklavik residents also considered the East Three area as poor fur and caribou country, preferring more familiar and productive grounds in the heart of the Delta and closer to home. But, complicating the situation, the social safety net provided by trapping began to disintegrate in the 1950s. In response to dwindling global fur prices, the Hudson's Bay Company closed many of its posts in northern settlements and converted others into retail stores. Native northerners had relied on these outlets throughout the year, not only to exchange furs for the staples of daily life but also to receive store credit if a season failed to yield a strong harvest. The federal government of Canada worried about the void created by the Company's decisions and sought to fill it through social services. The Family Allowance Act, for instance, provided funds to Native northerners to support the costs of child care. Eligibility for the allowance, though, required Native northerners to enroll their children in residential schools, which were operated in Inuvik. Inuit in the Delta, who had relied on mobility to mediate the arrival of whalers, expeditions, and reindeer, now felt stuck.[41]

When the government opened Inuvik over the end of 1959, some of Aklavik's 1,600 residents moved to the "model northern town," but many stayed. Those who refused to relocate saw their decision as bold self-determination and a form of resistance to colonial plans for the north. But they must also have realized the hard times they were in, without the fur economy that had for so long allowed them to navigate governmental interventions. Subsequently, Aklavik residents invented a motto for their home, "Never Say Die."[42]

MEGADEVELOPMENT AND ITS DISCONTENTS

Knowledge about permafrost helped US and Canadian leaders execute massive, complex industrial projects after World War II. By the late 1950s, the science of thermal balances and soil moisture had been applied to great effect, dotting a remote and "underdeveloped" region with wells, airfields, and "normal" living facilities. The Arctic was now more connected than ever to the south. Canadian officials, in particular, relied on this vision to separate themselves from a history of empire, whose legacies could still be found in the Arctic. When compared to the "New North" of Inuvik, the "Old North" now looked more like Aklavik, which had been "left to the missionaries [and] the fur traders," the mainstays of British imperialism. By pitting Aklavik as a problem and the "New North" of modern towns, oil economies, and permafrost scientists as the solution, outsiders implied that their interventions in the Arctic were somehow anticolonial.

This situation in Aklavik foreshadowed a set of power dynamics between Inuit and scientists that boiled over in the 1970s. When Aklavik residents decided not to move to the new town, government agents hired anthropologists and sociologists to investigate why. Academics and elected leaders treated this turn of events as a case study for modernization theory, reifying a boundary between indigenous and nonindigenous society that ran alongside the research of scientists. Inuit also noticed this boundary in their interpretations of modern history. According to interviews of Inuit living in the Mackenzie Delta in the early 1980s, the building of Inuvik forged a new generation of Inuit activists, who held a fundamental distrust of so-called experts. Inuit resented the notion that itinerant southerners, who spent perhaps a summer or two in the north, could "better detail the problems and potential of the area than its inhabitants." They also railed against the lack of effective communication and consultation made evident in governmental decisions about development. Indeed, after 1954, Inuit could easily connect the arrival of scientists with "some new governmental scheme."[43]

This same generation of Inuit leaders converted the creation of Inuvik and the associated expansion of defense and development into opportunities for resilience. Wage labor jobs—as mechanics, drivers, cooks, and carpenters—provided a steady supply of income and an exposure to new technologies, gains Inuit could no longer expect from the faltering fur economy. In the memories of Inuvialuit, the exclusion from megadevelopment after World War II—and involvement in it—paved the way for future political organizing, which eventually brought a century of colonialism in the Arctic to an end.[44]

CHAPTER FIVE

Disturbed: The Impacts of a Postcolonial Moment

It is practically impossible now to live off trapping only. That is the impact of all ways of transportation in our area: plane, helicopter, cat-trains on the tundra, seismic blasting on land and sea. Is this not a sufficient factor to disturb the animal life in land and sea?

. . . Too often in the past decisions and actions have been settled without consultation and we were faced with a matter of fact situation, unaware and unprepared . . . we believe that we should not only be observers, but participants and have a say in the deliberations.

CHARLIE GRUBEN, at the "Tundra Conference," in 1969, sponsored by the International Union for Conservation of Nature and Natural Resources

Charlie Gruben may have been standing at a podium that fall day at the Tundra Conference.[1] He also stood at the crossroads of history. Gruben became one the first Inuvialuit to speak before a high-profile meeting of Arctic researchers, conservationists, and North American policy makers. His invitation reflected a shared concern among Native northerners and a select group of southern power brokers about the impacts of energy development on Arctic environments. After the completion of Inuvik and Alaskan statehood, the federal governments of the United States and Canada leased millions of acres of the western Arctic to the oil industry. Multinational corporations bulldozed it, bombed it, and poked holes in it—all in search of oil and natural gas. Gruben put the ecological effects of these activities in a local perspective, but also in a humanist one: Big Oil was another example of outside intervention on Inuit lands. Over the 1970s, Gruben worked with scientists in the audience to lead Canadian Inuit to an agreement with federal officials for land title and fair representation in land-use decision making. Signed in 1984, the Inuvialuit Final Agreement enshrined Inuvialuit sovereignty and signaled the end of a century of colonialism in the western Arctic.

Literary scholars refer to an enduring "postcolonial" moment to define the period of time in which inhabitants of colonial spaces recreate attachments to the land, often through and against colonial structures. Of course, Inuit in Alaska and western Canada had adapted relations with nature in these ways since the arrival of whalers in the late 1800s. Gruben's speech confirms, however, the unique marginalization he and other Inuit experienced after World War II. Oil extraction obliterated the habitats of fur-bearing creatures, and thus the trapping lifestyle. Meanwhile, bureaucrats and industry representatives excluded indigenous peoples from northern administration. Gruben's presence at this conference also hints at the importance of Arctic scientists in campaigns for Inuit rights during the 1970s and 1980s. Just months after the gathering, and in response to public uproar over industrial pollution, the US and Canadian governments committed to evaluate the environmental impacts of development projects before approving them. This process, known generally as environmental impact assessment, created a space wherein scientists and Inuit concerned about Big Oil could advise government officials on its fate in the Arctic. Between 1970 and 1977, federal bodies from both nations conducted formal reviews of proposed pipelines, which would transport oil and natural gas found on the north slope of Alaska and western Canada to southern populations in the grips of a global energy crisis. It was clear to Gruben, then, that scientists—especially those in the "environmental" fields of biology and geology—now constituted an instrument of resilience within the very system of oppression that science had helped construct.[2]

When Inuit engaged scientists, they found some researchers who had begun to think about the Arctic in ways differently than their predecessors. Whereas geophysicists and engineers in the 1950s understood nature within a larger frame of national security and modernization, many biologists and geologists of the 1960s and 1970s saw Arctic phenomena through the lenses of ecology and environmentalism. Like Gruben, they had observed the oil industry's legacies on the northern landscape. They came to understand the Arctic as a system of interactions, many of which appeared uniquely fragile— and thus in need of careful oversight. These same scientists also disapproved of the ways governmental agencies traditionally conducted research—not only did they fail to incorporate concerns from impacted human communities, they also overlooked warnings their own staff provided of environmental damage. As a testament to the ways these critical scientific perspectives and Inuit voices aligned in this postcolonial moment, Gruben used the same word to describe Inuit society as biologists and permafrost scientists did to describe Arctic nature: disturbed.

In Arctic North America three profound historical trajectories converged on one another in the same instant: the extension of the petroleum economy, the elaboration of ecosystem ecology and mainstream environmentalism, and the expansion of an indigenous rights movement. Each of these forces shaped the other, and all left their mark on the northern environment of today. By 1984 federal governments in the United States and Canada had debated two Arctic pipelines, authorized one of these, and delayed the other because of its predicted environmental and social costs. They also settled land claims with all Inuit living in the western Arctic, from the north slope of Alaska along the Beaufort Sea coast into Canada. Rather than initiate a with-drawal of colonial forces—which might warrant the term "decolonization"— these events constituted another phase of the colonial encounter. In this postcolonial moment, scientists, governments, Inuit, corporations, and the land became only more entangled. A century of colonialism thus ends like it began: in a transition of power between established empires and rising na-tions, where science fostered the control of nature.[3]

BIG OIL ARRIVES

It was an aerial display, at first. Their engines piercing Arctic skies, fixed-wing aircraft transferred men and equipment to oil camps dispersed across the north slope of the continent. There, the ground show took over. Micheler sleds and tracked vehicles scoured the tundra for oil and then sucked the fields dry. In April 1960, one hundred million acres of the mainland Yukon and Northwest Territories were under oil and gas exploration. Across the border in Alaska, 1.4 million acres had been leased, with Texaco, Standard Oil, Mobil, Shell, and British Petroleum among the first on the scene. This was just a prelude to even more frenetic activity. In March 1968, Humble Oil and Atlantic Richfield discovered North America's largest oil field in Prud-hoe Bay, Alaska. By the end of 1971, the acreage of the Northwest Territories and Yukon under permit had doubled—to 464 million acres—while Alaskan leases quadrupled. In just ten years, entire swaths of the western Arctic had been converted into an oil frontier.[4]

It is difficult to underestimate the wide berth given to multinational oil companies, but easy to explain it. In other oil-rich parts of the continent, the high cost of crude owed to the multiple leases granted to companies over a single pool. The result was local overproduction and, in turn, pro-rationing of each well to meet market demand. In other words, the costs of oil were artificially boosted to make marginal wells profitable. The US and Canadian governments wanted to avoid this situation in the Arctic. They pursued bolder land use policies, which included wide well spacing and the direct avoidance of pro-rationing. The Alaska Statehood Act of 1959 allowed the state to select lands once owned by the federal government for its own jurisdiction, or to turn around and lease to oil companies. Meanwhile, in the Northwest Territories of Canada, all lands housing nonrenewable resources remained under federal control, despite devolution of some powers to the territorial government in Yellowknife during the 1960s. By controlling the process of land title and leasing, governments shaped an oil landscape that was similar to those in the Middle East and South America, rather than Texas and California.[5]

Between 1960 and 1967, oil companies spent roughly $25 million per year in exploration and development of wells in the western Arctic. By 1971, following the Prudhoe Bay strike, those average yearly expenditures ballooned to $167 million. Executives hoped that productive wells would be found, and cheap Arctic oil would soon compete with targeted markets in North America, Japan, and Europe. While crews drilled more than three hundred wells in the Yukon and Northwest Territories during the 1960s, only a handful of these became productive.[6] In many ways, exploration activities marked the arrival of Big Oil. The icon of the industry was the seismic crew, sent into the tundra with a geological map to locate hydrocarbons beneath the surface. In 1970 alone, seismic crews in the Arctic worked the equivalent of over 225 months of continuous labor. Given that a single seismic crew covered roughly 3 miles in a workday, crews surveyed more than 20,000 miles of tundra each year between 1960 and 1975.[7]

While technical, the operations of the seismic survey are crucial to understanding the environmental changes wrought by Big Oil and the consequences of these changes for science and colonialism. A seismic survey applied principles from geophysics to identify oil and gas formations invisible to the naked eye. The crew detonated a series of explosions on the ground, which sent a pulse of energy downward in all directions. These waves struck rock layers at various depths, bouncing back toward the surface. Because the subsurface environment is uneven, some waves traveled longer than others. To capture these differences, engineers measured the speed of returning waves with geophones. They converted these measurements into approximations of electrical energy and graphed them, which provided a two-dimensional version of the underground environment.[8]

On the Mackenzie Delta and north slope of Alaska, the most common type of seismic survey was the single line method. This method required the use of several tracked vehicles in a caravan, setting off blasts and collecting the data from them, and gashing vast stretches of the Arctic landscape. Such "seismic lines" are still visible from the air today. They are physical legacies of the ways multinational oil companies, governmental policies, and geological science combined to enroll Arctic nature into global energy economies. To those who know their full history, though, they are also a reminder of how ecological disturbance became a focal point for scientific and Inuit activism in the 1960s and 1970s.

SEISMIC LINES AND STORYLINES

The very scientists who had helped build the discipline of permafrost studies in the 1940s and 1950s were among the first to sound the alarm about seismic lines the 1960s. Permafrost scientists had established research programs in the Naval Petroleum Reserve in northwestern Alaska, and so had access to observe seismic lines before the expansion of Big Oil across the rest of the state and the Yukon and Northwest Territories. To these scientists, private enterprise was unbridled when compared to the discipline of the military. Geomorphologists working out of the Naval Arctic Research Laboratory were horrified to witness heavy machinery "ripping up the tundra" to build temporary roads or conduct seismic surveys. This was something the navy had tried early on, but learned would compromise the integrity of the permafrost below, rendering roads unusable within a season. One researcher wrote, "[they] were up to twenty years old and still getting wider! That was my first introduction to the fragility of permafrost." As this comment suggests, Arctic scientists pointed to seismic lines to understand an entire landscape succumbing to the petroleum economy.[9]

Geologists and geomorphologists linked what they already knew about the dynamics of permafrost with seismic lines to characterize the Arctic as easily disturbed. For example, physical geographer Ross Mackay studied a number of the "ubiquitous seismic line depressions" along the north slope of the Yukon Territory over the 1960s. He explained these as a product of "thermokarst subsidence"—the melting of the permafrost in the subsurface environment by the conduction of heat from the surface. Like permafrost scientists before him, he found that nearly any industrial activity on the tundra could alter the thermal balance, leading to slumping and sagging in the land. These activities included "the killing of vegetation without any removal of material, or the peeling off of the turf, or the compaction of peaty ground." These were the very same operations that corresponded to the work of the hundreds of seismic crews across the western Arctic (figure 13).[10]

As biologists joined geologists in studying a fragile Arctic, they highlighted permafrost's relationships within a broader system of Arctic life. The Arctic

FIGURE 13. With human figures for scale and a hill on the horizon as a vanishing point, Ross Mackay depicts the tundra disturbances wrought by seismic surveys along the Beaufort Sea coast in the Canadian Arctic. Such images constituted physical evidence for scientists of the fragility of the northern environment. Reprinted from Ross Mackay, "Disturbances to the Tundra and Forest Tundra Environment of the Western Arctic," *Canadian Geotechnical Journal* 7, no. 4 (1970): 425. Credit: © Canadian Science Publishing or its licensors.

appeared to have relatively few numbers of animal species—or, what we now might say, low biodiversity. As historian Stephen Bocking has shown, zoologists of the 1960s presumed low productivity conferred instability. The Arctic emerged as a special case where human activity could easily and quickly upset a natural balance. Mammalogists pointed out how nests and spawning areas, crucial for an animal's survival, were "particularly susceptible" to "permanent damage by man's activities." Botanists concluded that, depending on elevation and water content, vegetation communities were "unstable and readily effected by disturbance." On top of all this, plant life grew at remarkably slow rates, owing to portions of the winter without sunlight and relatively cooler temperatures the rest of the year. In his paper "Structure and Function of the Tundra Ecosystem at Barrow, Alaska," biologist Jerry Brown displayed these connections in a set of drawings and diagrams (figure 14). Brown showed how phenomena like "total soil nutrients" could be linked—whether hierarchically, or cyclically, depending on the line—to other phenomena, like species on the tundra grassland or the decomposition of animal matter. These storylines helped inculcate an ecological principle within scientific representations of the Arctic: everything is connected to everything else.[11]

For those scientists adhering to this ecological principle, the complex, fragile Arctic required coordinated scientific management. Frank Fraser Darling—a world-renowned animal behaviorist and popularizer of environmentalism—put it this way at the Tundra Conference in 1969: "How sensitive, how terribly sensitive, certain animals and biomes in the Arctic are." He continued, "The conservation-minded people and engineers are as one; one of their first principles is the conservation of the permafrost." Darling's declaration is significant, for at least two reasons. First, it positions the Arctic as a key site in the rise of ecosystem ecology and environmentalism in North America over the 1960s. Participants in the conference not only subscribed to precautionary principles as regards industrial development, they

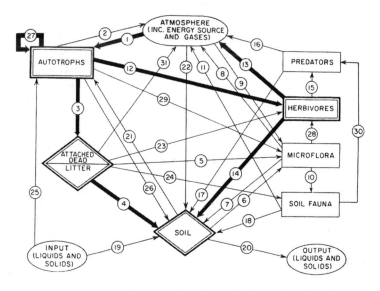

FIGURE 14. Jerry Brown provides an ecological storyline for the sensitive, disturbed tundra. The numbers stand for what Brown called "transfer functions." He hoped these could be quantified and loaded into a computer model, as a transition from conceptualizing Arctic environments to experimenting with their management, mathematically. Reprinted from W. A. Fuller and P. G. Kevan, eds., *Productivity and Conservation in Northern Circumpolar Lands: Proceedings of a Conference,* IUCN Publications, n.s., no. 16 (Morges, Switzerland: Unwin Brothers, 1970), 47. Credit: International Union for the Conservation of Nature.

also measured environmental change primarily through mathematical models of energy cycling. This was a sign of ecological science's evolution from early twentieth-century concepts like carrying capacity.[12] Darling's statement also identifies engineers—a group commonly villainized in accounts of popular environmental concern at the time—as contributors to, and supporters of, conservation in the Arctic. The arrival of Big Oil thus instigated a seismic shift in scientific orientation toward northern nature. In the 1950s, engineers and biologists recognized the profound effects of industrial operations on the tundra. In the face of military demands for defense and nationalist sentiments for development, however, they considered these effects as merely challenges to overcome with better research. Over the 1960s, a select group of critical biologists and geologists flipped this position: science ought to defend the Arctic from the disturbances of development.[13]

THE COEVOLUTION OF INUIT POLITICAL
IDENTITY AND SCIENTIFIC ACTIVISM

When Inuvialuk Charlie Gruben stood before conference-goers in 1969, he met scientists where they were—so to speak. By 1969, Inuit shared the language of a "disturbed" ecosystem that critical geologists and biologists elaborated over the preceding decade. But just as scientists had arrived at this perspective through careful study, Inuit grievances about oil development and tactics for airing them evolved over the 1960s.

Initial Inuit agitation against the oil industry drew from personal experiences with the military-industrial complex in the Arctic. For Inupiat living across the north slope of Alaska, a conflict with the US Atomic Energy Commission at Cape Thompson, Alaska, in 1958 became a touchstone when later confronting the expansion of oil economies. In response to the US government's plans to detonate an atomic bomb just outside the village of Point Hope, Inupiat leaped into action. Inupiat became enraged when the commission withdrew over 1,600 square miles of coastline for the experiment, without regard to indigenous title or prior Inupiat requests to claim the land under the Alaska Native Allotment Act. Native northerners gathered to share information about the proposal—code-named Project Chariot—raise concern about its potential impacts, and learn how such a project could be approved without their consultation. Referred to as *Inupiat Paitot*, or "people's heritage," the meeting forged the political organization of Alaska Natives—especially after the Atomic Energy Commission shelved Project Chariot in 1962.[14]

Over the 1960s, Inupiat Paitot remained a proving ground for Inupiaq activists in the fight against Big Oil and for sovereignty in Alaska. In 1962, the meeting grew considerably in size and spirit, featuring twenty-eight delegates from all Inupiat communities in the Alaskan Arctic. Inupiat voiced a set of common concerns: seismic lines were one, but so was the need for local control of economic development. The commonality among Inupiat experiences was remarkable, because of the geographic expanse between communities, the slight language differences among them, and the lack of transportation or communication networks—like roads, or television—to connect them. Indeed, efforts to bring together Inuit communities revealed the need for more regular information exchange among them and the scope of the environmental transformation in the Arctic. By 1964, Inupiat Paitot had spawned an organization with the same name, with bylaws and officers; circulated a newsletter, the *Tundra Times*; and organized a conference of indigenous people across the state. Between 1962 and 1967, Alaska Natives formed more than a dozen Native associations to reflect distinct regional indigenous communities of Alaska. The first of these, the Arctic Slope Native Association, started holding meetings in Barrow, Alaska, in 1965. One Alaska Native summed up the change in indigenous activism occurring through Inupiat Paitot over the 1960s: "Natives have never taken politics too seriously in the past," he said, but "politics can serve them if they pursue it wisely."[15]

As they developed this political identity in the 1960s, Alaskan Inuit deployed tools found in the administrative centers being built around them. These included lobbying organizations, but also knowledge about legal history. In a 1966 essay, William Hensley, an Inupiaq graduate student at the University of Alaska, argued that the Organic Act of 1884 provided the "basic protection of lands for Alaskan Natives." Section 8 of that legislation states: "Indians or other persons in said district shall not be *disturbed* in the possession of any lands actually *in their use or occupation* or now claimed by them" (emphasis added). For Hensley and other Alaska Natives, much of the controversy over oil exploration and Aboriginal title hung in these very words. Inupiat considered themselves occupiers of the entire landscape north of Brooks Range, given their subsistence livelihoods and history of trade relations. As Hensley put it, "the seasons, the movement of game, and the necessity of game conservation required periodic habitation and use of extensive areas." The Organic Act of 1884 thus provided a framework by which Alaskan Inuit challenged Big Oil and the state of Alaska. The imperative to document the historical occupancy of territory encouraged Inupiat to articulate their subsistence and stewardship in the Arctic as continuous over thousands of years. They cemented these historical ties through oral history and public testimony, but also through relationships with anthropologists and archaeologists. It was beyond question to Alaska Natives that Aboriginals had been "disturbed" in their possession of these lands. Nevertheless, the growing scientific record of the impacts of megadevelopment formed another component of the Inuit land claim.[16]

Indeed, Inupiat Paitot fostered close relations among Alaskan Inuit and ecosystem ecologists critical of the United States' military-industrial complex. As historian Daniel T. O'Neill has documented, a group of biologists hired by the Atomic Energy Commission to collect engineering data on Project Chariot became key allies for Inuit after finding startling data. Engineers behind the plan, like Edward Teller, had assumed that the site was unoccupied because of an absence of Native settlements there. But preliminary field research by Don Foote, William Pruitt Jr., Albert Johnson, and Leslie Viereck showed that Point Hope Inupiat hunted and fished there extensively. Importantly, this "Gang of Four" argued that the peculiar inner workings of the Arctic ecosystem guaranteed that these Inuit would be poisoned by radiation even if they were shielded from the initial atomic blast. Samples from lichens and sedges showed that nuclear fallout from bomb-testing over the Pacific Ocean in the 1950s had moved through the tundra with unusual efficiency. From these findings, the "Gang of Four" painted a picture of the Arctic food chain similar to what their contemporary Rachel Carson described in her groundbreaking *Silent Spring*. Unlike plants in southern climes, tundra plants drew the majority of their nutrition from the air. Caribou fed on lichens and little else. And Point Hope villagers relied on caribou for significant parts of their diet. If the plants at the base of these chains were contaminated, animals would concentrate the toxin, and Inupiat bodies would become radioactive.[17]

When the Atomic Energy Commission ignored these reports and failed to make them public, the "Gang of Four" were incensed. They took their conclusions to grassroots environmental groups in the southern United States. In 1961, Barry Commoner devoted a full edition of his popular antinuclear magazine, *Scientist and Citizen*, to the results of the Project Chariot biological studies. Especially because the project never came to fruition, Commoner later credited this episode as a milestone, when the American people recognized the audacity of the military-industrial complex and rose up in defiance of it. For Commoner, Project Chariot signaled a crack in the edifice of Big Science, that brand of research aligned with megadevelopment and wrapped in secrecy. Several Arctic scientists who had benefited from such arrangements at the Naval Arctic Research Laboratory in the 1950s—including Max Britton and Arthur Lachenbruch—were members of the committee on environmental studies. They enjoyed the support of the Atomic Energy Commission to conduct ecological research, but did not stand against the organization's plans like the "Gang of Four." There were, of course, professional risks associated with critiquing governmental plans for Arctic development. Following the censure of his studies of Project Chariot, Leslie Viereck resigned from the University of Alaska, writing, "A scientist's allegiance is first to truth and personal integrity and only secondarily to an organized group such as a university, a company, or a government." According to Pruitt, Inuit also rallied around biologists in their campaigns for environmental protection, Aboriginal title, and fair representation. The material in Commoner's bulletin, Pruitt said, "literally swept the Arctic coast, from Kakhtovik [*sic*] all the way down to Nome and below. . . . I recall, also meeting an Eskimo driving a dog team on the trail one time, and, by golly, he had a copy of [the issue] tucked inside his parka."[18]

In these ways, Inupiat and scientists organized around shared understandings of science, history, and justice. In 1966, three Inupiat in Barrow, Alaska, filed a land claim with the US government for the Arctic Slope Native Association. By the end of the year, more land in the state was under protest than the total number of acres in Alaska, because many indigenous territories overlapped. The influx of competing claims sparked a response from the secretary of the interior, Stewart Udall, who issued a land freeze on oil leases in the state until the issue of Aboriginal land title could be settled. As Inupiat had done, Udall justified his land freeze through the law of the land. "In the face of Federal guarantee that the Alaska Natives shall not be disturbed in the use and occupation of lands," he explained, "I could not in good conscience allow title to pass into other's hands."[19]

"MORE INVOLVED IN THE 'ACTION'"

Experiences in Canada followed the general pattern of Alaska. Like Inupiat, Inuit of the western Canadian Arctic first responded to the encroachment of government and industry in their homelands through direct resistance at the local level. Today, Inuvialuit recall walking off the job as cooks and equipment operators at oil campsites. Inuit in the Mackenzie Delta also later turned to technologies of communication—radio and newsletters—to air grievances, share information about the activities of companies across a vast landscape, and gradually form a unified indigenous political voice in Arctic

Canada. Like their neighbors in Alaska, Inuit in Canada faced a federal government that devolved some powers—in this case, to the territorial rather than state government—but nevertheless disregarded Aboriginal rights in their pursuit of northern development. Inuit also worked closely with scholars to document land use and occupancy in the terms of the legal record, which formed an essential component of their claim for land title. Interestingly, a biographical account of Nuligak, a Delta Inuk, circulated among local residents in the 1950s and popularized the story of Mackenzie Delta Inuit rejecting an offer of treaty from the Canadian federal government in the 1920s. Inuit used this manuscript as historical evidence that they had never dissolved Aboriginal title. Finally, Inuit in Canada directly engaged with biologists to document the disturbance of their homelands by the oil industry and win land freezes on oil exploration. A closer look at these partnerships with environmental scientists reveals they produced not only political and ideological alliances, but tensions, too.[20]

Inuvialuit relationships with activist-biologists emerged from an unlikely place, the Inuvik Research Laboratory. In 1964, the Department of Northern Affairs and National Resources formally opened Canada's first scientific station in the western Arctic, which had been transferred from Norman Wells. Officials in Ottawa, like Prime Minister John Diefenbaker, had hoped that scientists could use the building and its equipment to "modernize" northern social life and tease out the secrets of Arctic nature, especially oil and gas deposits. Meanwhile, Inuvialuit who relocated to Inuvik from Aklavik sought to direct the resource of the laboratory toward their interests. In 1968, Delta residents—including Elijah Menarik and Victor Allen—helped turn the building into something like a community college. Forming a group called the Mackenzie Institute, they occupied the building when the summer fieldworkers were away, opening its library and classrooms to the public. Allen, Menarik, and others scheduled programs oriented toward "the well being of northern residents," like managing businesses, training for northern research opportunities, and understanding government. In doing so, they transformed a scientific space, previously labeled as host of a "northern plague," into a vector of community organizing.[21]

Twelve months later, following the Prudhoe Bay strike in Alaska, oil companies descended on the Delta, ramping up surveying and drilling operations. During a period of haphazard seismic testing, where trap lines were destroyed and fish habitat disrupted, the Mackenzie Institute played mediator. Mackenzie Institute members consolidated research findings from scientists and distributed them to northerners in newsletters. Inuit partnered with biologists because they both were interested in where animals lived. Inuit had visited particular sites regularly and had witnessed change there over time—a kind of long-term monitoring biologists dream about. At the same time, the Mackenzie Institute collected observations of environmental change from Inuit who continued to live outside of town—dispersed in "bush camps" in the Delta landscape—and published these as well. In one of these newsletters, Inuvik resident Rose Mary Thrasher wrote that the Mackenzie Institute used radio reports, group discussions, and individual interviews to stay on top of the oil industry and the government, so that Delta residents could be "more involved in the 'action.'" Inuit had long negotiated their individual and collective futures through the trapping economy. With that livelihood crumbling beneath them, and with oil companies and federal governments failing to consult with them on Arctic development, science became a vehicle of mobility.[22]

While biologists could improve field research through these relationships, their motivations to participate in direct political organization are less clear. Even the adoption of an environmentalist ideology, which was widespread in professional biology in the late 1960s, does not suffice for an explanation. The Mackenzie Institute was clearly pro development. This would have put Dick Hill—the director of the Inuvik Research Laboratory, and a founding member of the institute—at odds with many of his conservation-minded colleagues at the Tundra Conference. A partial answer to this riddle can be found in the emerging trend of community development that swept through the Canadian federal government at the time the laboratory and the Mackenzie Institute opened.

The Department of Northern Affairs and National Resources (or Department)—the federal agency in Canada charged with northern administration—operated the lab through its Northern Coordination and Research Center. Since its establishment in 1955, the center had deployed research as a tool for ordering development across the Canadian north. It initially investigated the economic conditions of Inuit, to protect them from southern contact and incorporate them into the new mineral resource economy. In the early 1960s, though, the center forwarded a model of scientific engagement that was far more community driven, though no less paternal. The department dispatched a parade of development agents to northern settlements to build community councils. According to sociologists Frank Tester and Peter Kulchyski, these mechanisms were meant to control "the process of developing Inuit control" by directing northerners' concerns toward the municipal level, and away from the industrial economies being erected by government-sponsored private enterprise. Indeed, Canadian officials within the department were aware of the situation unfolding in Alaska with Inupiat Paitot and tracked the formation of Inuit rights organizations in Canada with unease. In this light, the presence of governmental scientists within the Mackenzie Institute might be seen as a means of infiltrating and derailing Inuit self-determination.[23]

There is another way to interpret the Mackenzie Institute, however. Dick Hill, who helped found the group, had been hired to run the Inuvik Research Laboratory by A. J. Kerr, the acting director of the Northern Coordination and Research Centre. In the estimation of geographers Peter Usher and Hugh Brody, who worked with Kerr in the 1960s, Kerr spearheaded a radical scientific program within the northern administration. He eschewed the old guard that believed in industrialization at all costs and inculcated in his staff a healthy skepticism of Ottawa bureaucracy. Kerr formulated this orientation toward science and public policy after logging twelve years in the Mackenzie Delta as principal of the governmental school in Aklavik. Like Inuit there, Kerr had been stunned by the conceit of engineers and administrators during the planning of Inuvik, especially their feigned interest in local concerns about land use. Thus when Kerr found himself in the position of shaping the trajectory of governmental-sponsored research in the Delta, he pushed it along a more just and socially accountable course.[24]

A politically engaged, humanist, and environmental science about the Canadian north flowered under Kerr's direction. Through the Research Centre, Kerr distributed monies to individual scientists and supported the initiation of northern studies programs at universities across Canada. He encouraged young researchers to stay in the field "as long as possible"—rather than become another summer pest—and pursue "honourable and equitable development" that prioritized the needs of northern peoples. Many of the reports published under Kerr's direction called into question the brand of large-scale resource development being promoted by the highest-ranking officials of the Department of Northern Affairs and National Resources. Many of the young scholars Kerr mentored went on to advocate for Aboriginal rights within budding Native political organizations, like Usher and Brody. These scholars remained thorns in the sides of senior scientists who refused to acknowledge indigenous interests. These same ethics informed Kerr's selection of Dick Hill to run the Inuvik Research Lab. When Hill asked a colleague why he was hired, Hill was told he provided "no pat answers about northern development" and had none of the familiar biases about Aboriginal issues. Put simply, while the Mackenzie Institute could appear as another sinister example of outside intervention, cloaked as community development, what matters more than its general organization is its internal composition. In the 1960s, a form of genuine critical science penetrated a technocratic northern administration. These scientists were hypothesis driven, but bent always toward social justice and ecological awareness. Like Inuvialuit, they wanted to be more involved in the action.[25]

Such scientific activism is borne out in Dick Hill's work with the Mackenzie Institute. Using his resources as director of the Inuvik Lab, he tapped a community of scientists and concerned citizens in North America to broadcast indigenous concerns about energy development. Writing to the *Arctic Circular*, a group of researchers, former civil servants, and upper-class urban residents, Hill asserted the importance of knowledge gathering for indigenous rights. He wrote, "The Mackenzie Institute contributes to total education so that more northerners may take their rightful place in social, cultural, and economic developments as citizens of the North, Canada, and the World." One can imagine Hill encouraging organizers of the Tundra Conference in 1969 to invite an Inuvialuk delegate like Charlie Gruben to speak before them about Aboriginal rights and oil activities in the Arctic. In these ways, politically organized Inuit and activist-scientists found each other, and began to shake up the established colonial apparatus from within.[26]

As Big Oil spilled over the western Canadian Arctic, Inuit and activist-scientists strategized effective protest. In January 1970, nineteen Native northerners formed the Committee for Original People's Entitlement (COPE), with several members of the Mackenzie Institute earning positions on the board of directors—including Charles Gruben.[27] COPE threatened Jean Chretien, the minister of the Department of Indian Affairs and Northern Development, with court action if more stringent protections on seismic surveys were not instituted immediately. Chretien flew to the Mackenzie Delta to meet with COPE members and, during the spring of that year, helped shepherd through the House of Commons important amendments to the Territorial Lands Act. Chretien committed to managing land use in the Delta according to knowledge about permafrost and tundra ecology, especially "the special susceptibility of some types of surface to permanent damage." According to the policy, industrial activities in the north would be licensed by the government of Canada according to site-specific conditions—for instance, the water content of soils or the presence of bird nests. Land users would have to "account for damage done to the land," though it was not made clear what mechanisms would ensure this accountability. One of the scientists responsible for drafting these directives was biologist William Pruitt Jr., who had earned his stripes as an environmental scientist and activist—one mindful of indigenous rights—while working with Inupiat in Alaska to combat the US Atomic Energy Commission's Project Chariot.[28]

These arrangements for land use planning in the Arctic represented an important victory for activist-scientists and Inuvialuit. Both were thus sorely disappointed when the northern administration failed to implement them fully. The summer of 1970 witnessed continued seismic blasting in the Mackenzie Delta, with no consultation of Inuit and little oversight by government land use inspectors. In July 1970, COPE helped organize the Coppermine Conference, a meeting not unlike Inupiat Paitot in its reach: thirty Inuit representing more than twenty different communities across the Canadian north, from the Mackenzie Delta region to northern Quebec, descended on the town of Coppermine (now Kugluktuk). Conference-goers made their voice heard in a telegraph to Prime Minister Pierre Trudeau. The western Arctic, they wrote, "has a very delicate ecological balance which will be destroyed if exploration is allowed and the land surface disturbed." The Canadian government could easily understand this, Inuit continued, "it if will only choose to consult with the Eskimo people and have biologists conduct the necessary research." In these ways, the Coppermine Conference of 1970 reflected the decade-long coevolution of Inuit political identity and a critical, humanist, and environmental science.[29]

ENVIRONMENTAL IMPACT ASSESSMENT ARRIVES

And then, suddenly, everything changed. Responding to what we now call mainstream environmentalism, the US and Canadian governments institutionalized environmental science and public comment periods in the design stage of major energy projects. When environmental impact assessment arrived in the Arctic in the early 1970s, it brought an open, untested political framework for organized indigenous and scientific activism.

On New Year's Day 1970, President Richard Nixon signed into law the National Environmental Policy Act (NEPA). The act instituted procedural requirements for all development on federal lands or initiated by federal agencies. Before authorizing any project, relevant federal bodies had to prepare and review an "Environmental Impact Statement," an evaluation of the predicted environmental effects of a proposed action. The government of Canada also experimented with environmental impact assessment tools in the early 1970s. It did not adopt a legislative approach out of fear that it would induce litigation, as it had begun to do in the United States. Just months after passing NEPA, US scientists produced over four thousand Environmental

Impact Statements, many of which faced legal challenges from conservation organizations who claimed they had not been prepared according to the law. One of these challenges, upheld by the Supreme Court, mandated that environmental impact statements fully disclose the risks of proposed actions, and that relevant governmental agencies increase public participation in the decision-making process through open comment periods.[30]

Canada worried that legislating environmental impact assessment procedures might shift authority over land use from government agencies to the courts. To prevent this, the Canadian government mandated impact assessment through cabinet orders instead. By the end of 1973, the government established guidelines for environmental impact assessment, which extended a "self-assessment" standard. In other words, in Canada, the industries proposing development conducted their own impact studies and submitted them to federal authorities responsible for approval. The government also created an independent review board to evaluate impact statements developed by project proponents and make recommendations to the federal agency with authorization powers.[31]

The National Environmental Policy Act of 1970 in the United States and cabinet mandates in Canada overhauled the existing political framework for managing the Arctic environment. Importantly, decision making had to be transparent, and governmental planning had to at least incorporate the interests of the people, if not serve them. Scientists had experience navigating the interests of administrators and industry representatives, but such research was rarely, if ever, subjected to public scrutiny in the age of the academic-military-industrial complex. Even more, environmental science now had to be predictive rather than reactive—it had to anticipate and evaluate the consequences of development and not merely record its effects. These politics of science in the 1970s differed substantially from the mission-oriented research during both megadevelopment and community development in the previous three decades. But they opened a door for activist-scientists whose commitments to ecology, environmentalism, and social justice aligned with the principles built in to the idea and structures of impact assessment.

For their part, oil industry representatives seemed ready to take on the shifting roles and processes of decision making, especially when it came to Arctic pipelines. At a conference held at Harvard University in October 1970, Inuvialuit from Canada's western Arctic sat alongside biologists, governmental officials, and the president of Atlantic Richfield—the oil company responsible for discovering the largest oil field in North America at Prudhoe Bay, Alaska. Writing in the conference proceedings, company executives admitted the misgivings and consequences of oil exploration in the 1960s.

Emphasizing "smart development," organizers hoped "to bring about a dialogue between the whole spectrum of disciplines and sectors of society that have a stake in the Arctic," while establishing the importance of "industrial management" in northern affairs. When oil companies did seek approval for Arctic pipelines in the early 1970s, procedures for environmental impact assessment produced this very dialogue—but perhaps not the consequences that anyone had in mind.[32]

THE TRANS-ALASKA PIPELINE PROPOSAL AND THE ALASKA NATIVE CLAIMS SETTLEMENT ACT

In February 1969, the Trans-Alaska Pipeline System—a joint group created by Atlantic Richfield, British Petroleum, and Humble Oil—requested permission from the US Department of the Interior to construct a pipeline from Prudhoe Bay, Alaska, to Valdez, 800 miles to the south. In 1974, after several years of drafting and commenting on environmental impact statements, and in the midst of a crushing energy crisis, the US government approved the Trans-Alaska Pipeline System. Scientists felt that while the assessment process allowed for the synthesis and dissemination of a generation of research on the Arctic, the ambiguity of the existing National Environmental Policy Act allowed economic interests to dictate the final outcome. Inupiat were trapped between their concerns over oil exploration—which would take place largely on their lands—and their advocacy with fellow Alaska Natives for Aboriginal land claims, which were partially recognized in the Alaska Native Claims Settlement Act of 1971. In the end, Inupiat earned land title through this legislation, but they did not earn self-determination. This initial test of the National Environmental Policy Act thus produced a string of effects—both intended and unintended—that continue to shape the Arctic today.

To see how this could be so, we must first inspect the process of decision making that the National Environmental Policy Act spawned in Arctic Alaska. In response to the application for a right-of-way, the US government formed the North Slope Task Force, a committee with representation from a slew of governmental agencies. In order to examine the environmental impacts of the proposed pipeline across great stretches of the Alaska landscape, the North Slope Task Force created a subcommittee, itself housed at the Alaska Branch of the US Geological Survey's office in Menlo Park, California. By design, this subcommittee included many scientists who had studied permafrost during the 1940s and 1950s at the Naval Arctic Research Laboratory and in the Naval Petroleum Reserve in northwestern Alaska.[33]

In assigning permafrost scientists to review Arctic pipeline proposals, the North Slope Task Force created a venue within the public domain for the circulation of scientific knowledge gained since World War II. Through this channel, permafrost scientists and biologists rebroadcast concerns they had with development in the Arctic. Researchers noted that the proposal for the Trans-Alaska Pipeline System had not fully considered the revegetation of disturbed areas, the need to protect streambeds and fish spawning areas, or the possible restrictions on the free movement of wildlife posed by the pipeline itself. Scientists pointed directly to the project proponents' ignorance of the sensitivities of permafrost, since a buried route was their preferred means of transporting oil from the north slope. They also identified a less obvious environmental impact: the "millions of cubic yards" of gravel that would be required to insulate the ground around the sections of pipe to be laid below the surface.[34]

Meanwhile, as scientists worked within the impact assessment process, Alaska Natives found themselves absorbed by the federal government's process of settling land claims. Both the state of Alaska and the Department of the Interior considered Aboriginal title a barrier to energy infrastructure and committed with urgency to making a deal with Alaska Natives. As the language of the Alaska Native Claims Settlement Act makes clear, federal officials consciously shaped the agenda away from self-governance and toward land title. The United States explicitly avoided "any permanent racially defined institutions, rights, privileges, or obligations" as well as the possibilities of a "reservation system or lengthy wardship or trusteeship." While such an approach avoided paternalism, it also constrained possible architectures for indigenous self-determination. The federal government treated all indigenous claims together and wanted any act of Congress to terminate future relations with Alaska Natives. This became an issue for Inupiat, who, because of geography, held a unique position relative to other indigenous Alaskans as regards oil exploitation. The government initially proposed an arrangement whereby cash and land would be transferred from the state and federal holdings according to the size of the population of a Native association, giving more money and property to regions with more people. The Alaska Federation of Natives got on board with this, but Inupiat disagreed, for several reasons.[35]

The Arctic Slope Native Association's claim was the largest, geographically speaking, but had one of the smallest populations. Inupiat argued that the value of the land—not just its size—should be considered. To quantify this value, and see the disparity between the north slope and other regions in Alaska, Alaska Natives need look only to recent land leases by the state. With interest and appetite for domestic supplies of oil growing at the end of 1969, the state of Alaska earned $900 million from bids on 450,000 square miles of land adjacent to Prudhoe Bay. This dwarfed the $12 million in sales they had made between 1965 and 1967. Inupiat communities decided to leave the Alaska Federation of Natives in protest because the effects of oil development were disproportional to the rewards of sovereignty. After heated discussion, Inupiat rejoined the federation, once that group revised its plan such that any monies from Congress would be divided according to lands lost in the extinguishment of title. The federation moved to accept a settlement with the federal government on these terms. There were but fifty-six dissenting votes, all from Inupiat.[36]

In 1971, President Richard Nixon signed the Alaska Native Claims Settlement Act, the largest settlement of its kind in US history. The act transferred more than $900 million and 40 million acres of land to Alaska Natives, in exchange for extinguishing Aboriginal claims to an additional 300 million acres of Alaska, as well as subsistence hunting and fishing rights. The indigenous land was divided between two hundred "village corporations"—which corresponded with existing Native settlements—and twelve "regional corporations," formed on the basis of historic and cultural compatibilities that connected discrete settlements. The Northwest Arctic Native Association (now NANA Regional Corporation) and the Arctic Slope Regional Corporation officially formed in 1972, constituting the only corporations in Alaska's Arctic territory. The NANA regional corporation now owns more than two million acres of land in the Noatak, Kobuk, and Selawik river basins. The Arctic Slope Regional Corporation oversees five million acres lying north of the Brooks Range. Yet these properties are less than 10 percent of the area that Inupiat called home. The Settlement Act also dissolved Aboriginal title to more than 56.5 million additional acres in Arctic Alaska—a swath of foothills, coastal plain, and deltas that drained into Kotzebue Sound and stretched from the Brooks Range to the sea. Land claims thus redrew the boundaries of jurisdiction in the north, in ways that both legitimized Inuit sovereignty and circumscribed it. Inupiat knew the ecological impacts from industrial activity respected no political boundaries, but could exercise little authority beyond their territories and corporate structures.[37]

As became clear to Inupiat living in Arctic Alaska after 1971, fighting for ownership of the land was not equivalent to establishing fair practices for deciding how that land ought to be managed. Following the Alaska Native Claims Settlement Act, the federal government allowed Inupiat in the Arctic Slope Regional Corporation to choose 10 percent of state lands for its own jurisdiction. Inupiat focused their land selections on areas around Prudhoe Bay, known to have significant deposits of gravel—the crushed rock found in isolated pockets beneath the tundra and used in all types of construction. The state of Alaska and the federal government balked at this decision and then blocked it. Transferring these gravel deposits to Inupiat hands would have garnered them serious bargaining power with multinational corporations developing oil resources and building infrastructure—like pipelines— on state or federal lands.[38]

Meanwhile, the processes for decision making unfolding before them— environmental impact assessment and land claims settlement—also limited the ways Inupiat could shape land management. When the final Environmental Impact Statement was made available for public comment, the US government narrowly interpreted the term "public." Seven copies of the volume were distributed to select governmental offices across the country, for a period of forty-five days, and opened for reading or comment only during business hours. Needless to say, Inupiat living in northern communities were at a disadvantage in providing input. An appeal to Congress for more time and more extensive public access was denied. As Billy Neakok recalled, Inupiat were disappointed by both the land selection and public participation processes, as they found their efforts to "assert local control over the environment" thwarted.[39]

The completion of the impact assessment of the Trans-Alaska Pipeline System was a mixed bag for Arctic scientists as well. As historian Peter Coates has detailed, the National Environmental Policy Act's language at the time did not specify how federal agencies should authorize development. Was an Environmental Impact Statement enough for development to proceed? Would a project be aborted if scientists or the public considered the impacts intolerable? Federal officials mulled these uncertainties as they watched the oil industry, the state of Alaska, and the Alaska Federation of Natives align in support of the pipeline. Then in the summer of 1973, an oil embargo pushed prices at the pump to unprecedented levels in the United States. Taking swift action, Nixon signed the Trans-Alaska Pipeline Authorization Act in November of that year.[40]

As construction on the pipeline began in March 1974, Inupiat must have felt shocked by the changes in the political and physical landscape. Just like oil exploration during the 1940s and 1950s, the Trans-Alaska Pipeline

System became the largest megadevelopment in Alaskan history. Whereas defense construction in the 1950s required 7.34 million cubic meters of gravel, Big Oil and its contractors needed ten times that amount for the building of the work pad alone. This was another painful reminder for Inupiat that land title alone was no guarantee of the kind of autonomy they desired.[41]

THE MACKENZIE VALLEY PIPELINE INQUIRY AND THE INUVIALUIT FINAL AGREEMENT

In March 1974, the same month that construction began on the Trans-Alaska Pipeline System, oil companies applied for a right-of-way for a natural gas pipeline to cross the north slope of the Yukon Territory, enter the Mackenzie Delta, and follow the Mackenzie River south to Alberta. The Canadian government promptly tasked Supreme Court justice Thomas Berger to inquire into the "terms and conditions" of such a pipeline in light of its potential social and environmental impacts, as well as means of mitigating or avoiding them. Just as the framework of the Trans-Alaska Pipeline proposal frustrated scientists, ushered in a pipeline for Arctic Alaska, and presented challenges to Inupiat sovereignty, impact assessment created new possibilities in Canada and foreclosed upon others.[42]

The Mackenzie Valley Inquiry, as it came to be known, ultimately recommended that oil development could not proceed until land claims were settled with Aboriginal communities living in the pipeline's proposed route. Thomas Berger asked for a ten-year moratorium on construction for these settlements to be worked out. In addition, he argued that no energy corridor ever cross the outer Mackenzie Delta, because of the unique ecosystem there. Berger demanded, "special measures" to avoid "disturbance to fish populations" and harassment of migratory waterfowl and beluga whales. Seven years after the completion of the inquiry, Inuvialuit residents of the Mackenzie Delta region and the government of Canada signed the 1984 Inuvialuit Final Agreement, the first comprehensive land claims agreement in the Canadian north.[43]

Why did Berger take these stances? Many scholars point to Berger's leanings toward indigenous and environmental issues—which he had already made clear in a 1973 landmark case for Aboriginal title. Close readings of the testimony of northerners at the inquiry hearings conducted across the north show that Inuit were not antidevelopment, but rather rejected oil production on their lands if they had neither title to land nor say in decisions about it. Historians and political scientists have also noted the alliance among environmental and indigenous lobbies, who fought the approval of Arctic pipelines. Yet there were other forces at play that deserve recognition and make visible the agency of scientific representations in this postcolonial moment. Certainly, Berger made recommendations that privileged Aboriginal title over the application of the best science available. But all three parties—researchers, Arctic residents, and Berger himself—were of the same mind on another issue: taking care to treat the Mackenzie Delta as a fragile and disturbed environment.[44]

Berger noted how Delta residents helped him arrive at this conclusion. In his final report to the Canadian government, he documented his visit to Archie Headpoint's camp outside of Aklavik, in the middle of the mazelike Delta, as a source for his recommendations for federal protections in the area. Sitting in a "collection of small, cluttered buildings," resting just above the bank of the Mackenzie River, Berger and Headpoint could see "one of Shell's seismic exploration camps" on the iced-over waterway. But they need not look that far to see how the oil industry was changing the Delta. In the "few acres around that cabin," Berger noted the same scars on the tundra that many before him had studied and grew concerned about. "There, the landscape is crisscrossed by seismic trails and vehicle tracks that seem to come from nowhere and go to nowhere," he wrote. Like Inuvialuit and biologists, Berger connected these trails to their knock-on effects throughout the ecosystem, up the food chain to Native northerners. "The Headpoints complained that the land was no longer as productive as it had been," Berger recalled. "The seismic trails . . . had blocked the streams and polluted the ponds."[45]

Because of his history dealing with issues of Aboriginal title before the Supreme Court, Berger may have been predisposed to see the links among ecology, commerce, and the social conditions of trappers. Regardless of his outlook, though, Berger was required to make these links because of the terms of reference for his inquiry. In 1970, four years before the inquiry, the minister of Indian Affairs and Northern Development sent a reconnaissance team of scientists up the Mackenzie River to begin collecting data to help facilitate land use planning in the region. Called the Mackenzie Delta Task Force, this team identified the basic criteria upon which land use permits could be authorized for developers in the Northwest Territories. The Task Force—and the Arctic Land Use Research Program that it eventually became—would be the government of Canada's agency to evaluate the integrity of industry's environmental impact statements. Importantly, surveys by the Task Force in the Delta in 1970 provided an opportunity for Native northerners and activist-scientists to influence nascent governmental environmental policy. While in the Delta, scientists met with several Native leaders, who argued that land use permits should be approved not only on the basis of impacts to biota and permafrost but also in accordance to disruptions of fishing, trapping, and hunting. In their report to the Department of Indian Affairs and Northern Development, the Task Force noted that, "any evaluation of how exploration may be affecting [Native] hunting, trapping, and community affairs" fell within the scope of the new regulations.[46]

When Thomas Berger determined the conditions necessary to build the Mackenzie Valley Pipeline in the mid-1970s, he relied on this existing northern land use policy. "I have proceeded on the assumption," he recalled as he communicated his decision to the minister of Indian Affairs and Northern Development "that we intend to protect and preserve Canada's northern environment, and that, above all else, we intend to honour the legitimate claims and aspirations of the native people." For Berger, that charge now constituted the framework within which he understood the threshold of environmental impact. "All of these assumptions," he wrote, "are embedded in the federal government's expressed northern policy for the 1970s."[47]

Still, simply because scientists and governmental adjudicators had to account for subsistence livelihoods in their ecological research on proposed pipelines does not mean that the various stakeholders involved would agree about the scale of environmental impacts in the Mackenzie Delta. Yet in testimony before Thomas Berger, Inuit and scientists—whether hired by industry or employed by the Canadian government—often did speak to the same effects of the proposed pipeline in this place in similar ways. Consider the exhibit of Inuvialuk Bertram Pokiak, during a community hearing administered by Berger in the hamlet of Tuktoyaktuk in March 1976. Pokiak highlighted three sensitive regions within the Delta: Husky Lakes, Kugmallit Bay, and Kendall Island. On one hand, Pokiak zeroed in on these areas because Inuvialuit frequented them during subsistence activities. "Virtually the entire population of Tuktoyaktuk" fished in Husky Lakes "every spring," he said, while other sections of the area were used for "fishing, trapping, and hunting caribou and moose." Kugmallit Bay remained the central summer whaling camp "where most of the Tuktoyaktuk whale hunters congregate," while Kendall Island received goose hunters in April and May. On the other hand, Pokiak and other Inuvialuit highlighted these areas because the animals found in them were particularly disturbed by industrial activity. Previous seismic surveys at Husky Lakes had "blocked spawning runs" for fish. The logistics of delivering supplies via the Beaufort Sea and Mackenzie River had scared belugas during their migration to Kugmallit Bay. And "low flying aircraft" during the spring migration of waterfowl were "interfering with the birds' movements." Pokiak's recommendations for these locations matched exactly the recommendations by the Environment Protection Board—the group of scientists assembled by one of the project proponents, Canadian Arctic Gas, to draft environmental impact statements. The scientists Thomas Berger retained to analyze Big Oil's science also focused on these same locations, as evidenced in Berger's final report when he advocated for bird and whale sanctuaries in the Mackenzie Delta.[48]

This convergence owes in part to the politics of environmental assessment in Canada at the time—particularly, the logistics of collecting data on impacts and analyzing it in a manner sufficient to federal governments. The oil companies operating in the Canadian Arctic in the early 1970s were largely the same as those that had proposed the Trans-Alaska Pipeline in 1969. They understood that impact statements were not only necessary for approval of right-of-way, but the information they contained could also enhance business practices. There were cost savings in streamlining construction and in reducing the likelihood of a costly spill during operation. Oil consortia working in Arctic Canada thus went to great lengths to ensure rigorous science. They brought on board many researchers who had experience in the Alaskan pipeline process and the Project Chariot environmental studies to facilitate field research and final reports. At a conference in 1977 devoted to improving the research side of impact assessment, one of these scientists—A. W. F. Banfield—reflected on the opportunities oil companies provided to develop the relatively new field of impact prediction. Arctic Gas and Foothills, Ltd. gave scientists more than $4 million to create experimental facilities to test permafrost thaw, simulate disturbances of wildlife by helicopters and other machinery, and run computer modeling and statistical analyses to determine if certain human activities had greater impact than others. Engineers, biologists, and geologists also collaborated on a series of map overlays that showed conflicts between spawning areas, permafrost zones, and proposed pipeline routes, such that construction designs could be altered to reduce impacts. The scientific community and oil executives thus began to view the effects of the pipeline in similar ways, even if this meant acknowledging how a pipeline would change Arctic nature. Thomas Berger himself said as much during the inquiry. He noted that Arctic Gas and Foothills, Ltd. "did not seek in any way seek to interfere" when biologists testified that the project was "not environmentally acceptable."[49]

Inuit in the Mackenzie Delta also circulated this newfound knowledge about pipeline impacts, completing the convergence among residents of the western Arctic, industry representatives, and hired scientists. This circulation followed two routes. First, field biologists built a rapport with Inuit during the collection of preliminary data in the first four years of the 1970s. Biologists engaged local residents, whether at research sites in the Delta, at meetings of Hunters and Trappers Committees in town, or at the Inuvik Research Laboratory, where some Native northerners found employment as technicians. Second, once project proponents published final reports in 1974, Inuvialuit political organizations shared them with Inuvialuit across the Delta and Beaufort Sea coast. Like the Mackenzie Institute did in the late 1960s, the Committee for Original People's Entitlement conducted more than 24,000 interviews in the region in the twenty-four-month period between the release of the environmental impact statements and Berger's community hearings there. The committee's field-workers translated for elders the pipeline's expected effects on flora, population densities, habitats, and more. Taken together, these pathways of knowledge sharing fostered a common understanding across industry, government, and Inuit organizations about the specific impacts of a pipeline in the Mackenzie Delta. That these stakeholders shared a language in describing the scale and significance of environmental impact was no accident.[50]

When Berger made his recommendations about the pipeline, Inuvialuit rejoiced. But not all biologists were as happy. Larry Bliss and R. D. Jakimchuk, who helped compile Arctic Gas's impact statement, thought the unprecedented financial and institutional support for impact prediction had helped advance the state of knowledge about Arctic nature. They interpreted Berger's decision as an indictment of the science of assessment itself. Bliss felt that the gains made by scientists were lost when Berger recommended against a pipeline, rather than setting out the terms on which it could be built. In other words, if one of the longest, most rigorous, and most interdisciplinary research programs ever to take place in the Arctic could not provide definitive solutions for avoiding and mitigating impacts, which one could? Bliss and Jakimchuk also showed frustration about the adversarial nature of assessment proceedings, in which a scientist's assumptions and even personal motivations could be called to question.[51]

While some chafed against decision-making procedures that "exposed the cloistered discipline of biology," others embraced them. These other scientists, like Thomas Owen, pointed to the inquiry as highlighting the need for "fundamental changes in the way that knowledge is organized, transmitted, and applied" in a society increasingly concerned with the "future consequences of today's actions." Owen noticed how the inquiry initiated a shift in the persona of academic experts from "objective" observers to "science advocates." Scientists were forced to move debates about methodologies and conclusions from technical journals and private conferences into the public realm. There they took up the controversial subject of making value judgments on their data—what was the "best" way to understand an impact, and when was an impact not "acceptable"? The inquiry pushed science even further "upward and outward" to a wider section of society for deliberation. Ultimately, Owen realized, impact assessment created a new kind of science, one he called "postindustrial." This version was inherently interdisciplinary, problem oriented, and vertically organized—where specialists not only communicated across disciplinary lines but also with interest groups and individual citizens "whose right to participate is guaranteed by law." By the late 1970s, the crack in Big Science first noticed during Project Chariot had undermined the foundations of knowledge and power. Impact assessment was a means of rebuilding more equitable and rigorous foundations.[52]

Seven years after Berger's decision on the Mackenzie Valley Pipeline Inquiry, Inuvialuit and Canada signed the Inuvialuit Final Agreement. Like the settlement in Alaska, this one made history. It was the first comprehensive land claim agreement signed north of the 60th parallel and only the second in Canada at the time. It established the sovereign territory of the Inuit of Canada's western Arctic through a transfer of 35,000 square miles of surface and subsurface land rights, while extinguishing Aboriginal title beyond this landmass. Inuvialuit secured sites with gravel, affording them crucial bargaining chips in future oil development. They also selected lands from the federal government based on scientific studies of their ecological sensitivity, biological productivity, and mineral resources. Notably, the Inuvialuit Final Agreement included provisions not found in the Alaska Native Claims Settlement Act. Inuvialuit wrested from federal agencies some control over Arctic land use decisions—whether wildlife, fisheries, or development proposals— through guaranteed involvement with governmental officials on comanagement boards and environmental impact assessment bodies.[53]

THE POSTCOLONIAL STRUCTURES OF SCIENCE IN THE ARCTIC

Importantly, that Inuvialuit looked to the control of impact assessment as means of self-determination was itself a product of Inupiat failures to achieve these same mechanisms with the US government. In these ways, science and the Arctic differ across the international border today, even as the place remains internally interconnected and linked with the postcolonial moment of the 1970s and 1980s.

The authorization of the Trans-Alaska Pipeline System and the limitations of the Alaska Native Claims Settlement Act forced Inupiat to get creative in their pursuit of self-governance. State law in Alaska allowed for the creation of a "borough" unit that allowed its members to pass resolutions on neighboring state and federal lands. Through zoning, Inupiat could protect lands used for subsistence by forbidding development there. Through taxation, Inupiat could gain revenues from oil industry activities to fund initiatives for self-government, education, public health, and more. In 1972, Inupiat applied for borough status for the entire territory north of the Brooks Range, prompting the state of Alaska to determine whether or not that land was used by Natives, and thus could be considered a borough. With its creation, the North Slope Borough became the largest municipality in the world. Roughly the size of California, Washington, and Oregon combined, it borders two seas and two nations. In 1986, Inupiat in villages around the Kotzebue Sound followed the lead of the North Slope Borough, forcing the state of Alaska to form and recognize the Northwest Arctic Borough. By the end of the decade, the Arctic Slope Regional Corporation and the NANA Regional Corporation were some of the most successful indigenous corporations in Alaska. Both employed high percentages of beneficiaries in Native-owned businesses, paid dividends to shareholders from the development of natural resources, and turned cumulative profits—a feat that only three of the thirteen Native corporations accomplished.[54]

The establishment of borough units partially restored an Inupiat land base that they had negotiated until the arrival of commercial whalers in the late 1800s. Importantly, Inupiat now mediated governmental and corporate interest in that land not only through policies that taxed oil companies, but also through scientific research. In the years following its creation, the North Slope Borough witnessed the continued development of oil resources, at the Pet 4 Petroleum Reserve and in the Prudhoe Bay fields. Yet Inupiat had not been included in environmental impact assessments in these activities. In 1976, Eben Hopson, the first mayor of the North Slope Borough, formed the Department of Conservation and Environmental Security within the municipality to draw attention to this procedural injustice. He appeared before the President's Council on Environmental Quality to argue that the federal government had neglected its national duty to the Arctic environment by disregarding Inuit governmental bodies in the review of draft environmental impact statements. He and other Inupiat also began reaching out directly to Arctic scientists, to create a process to evaluate impact statements that incorporated Inuit voices. Hopson articulated the value of bridging native knowledge and scientific research. Speaking to a group of Arctic researchers, he noted, "For, in our language is a whole natural science of the Arctic, and we feel that others should listen to us when we warn them against making a serious mistake." But Hopson went further than advocating for better knowledge production through recognition of different ways of knowing. He also pushed for improvements in environmental impact assessment, what he called the "undisciplined discipline." "Through a sharing relationship," he charged, "we could enable faster, environmentally safer, Arctic resource development."[55]

Hopson found in science and impact assessment a lever to wrench authority over land management. This was the kind of power Inupiat once held, but which had slipped away with increasing speed after World War II and formally stripped away with statehood and the Alaska Native Claims Settlement Act. Through the creation of their own agencies within boroughs, and close collaboration with scientists, Inupiat resolved to shape impact assessment toward their interests. Hopson went further in his promotion of Inuit rights.

With funds from the borough, Hopson held the Inuit Circumpolar Conference in Barrow in 1977, a meeting of Inuit peoples from Greenland, the United States, Canada, and Russia that produced a lasting international organization of the same name (now Inuit Circumpolar Council, or ICC). Over the last four decades, ICC has asserted the rights of indigenous Arctic peoples to self-determination, economic development, and environmental security. Hopson pushed for international cooperation because of what he saw playing out with his Inuvialuit neighbors, who were in the midst of the Mackenzie Valley Pipeline Inquiry. He worried that the Canadian government and the scientists hired to study the impacts of the proposed pipeline had too narrowly framed the question of oil development in the Arctic. He noted that while the pipeline would be on shore, drilling would take place off shore. If an oil spill or gas blowout occurred, ocean currents would circulate pollutants to the north slope of Alaska, threatening livelihoods and ecologies in the Inupiat settlement area. To these ends, Hopson built partnerships among Inuit peoples and scientists on both sides of the international boundary in northwestern North America. Beyond the Inuit Circumpolar Conference, he targeted the Alaskan Science Conference. In a 1976 address to the group, he invited scientists to commit to more community-based practices, where researchers and residents acknowledged each other's unique expertise to organize "a comprehensive regional circumpolar science agenda." He saw cooperation between scientists and western Arctic Inuit communities as creating a single "laboratory," which was free "from the limitations imposed by narrow and specialized national and industrial interests." He also authorized a loan from Inupiat to Inuvialuit to help hire legal and scientific staff for Inuvialuit dealings with the Canadian government. These funds, drawn from the cash payout of the Alaskan Native Claims Settlement Act of 1971, went toward hiring lawyers as well as Bob DeLury, a biologist and chief negotiator for the Inuvialuit during land claims discussions. With these exchanges, Inuit from across borders united to shape Arctic nature and Arctic governance in ways reminiscent of trade rendezvous at Kittigaaryuit one hundred years earlier.[56]

During negotiations with Canada, Inuvialuit were not satisfied with land title alone. They wanted to ensure consultative involvement in all decision making in their settlement region. One of the negotiators for the Inuvialuit demanded that any agreement include indigenous participation in the "review of applications for land use permits," as well as "more technical assistance" from scientists. Canada agreed to this in the Inuvialuit Final Agreement, and over the early 1980s, representatives from both Inuit and federal governments began implementing such procedures. The result is a set of boards that oversees fisheries and wildlife management in the Inuvialuit Settlement Region, and also handles development proposals, with administrative and scientific support from the federal government. According to the agreement, Inuvialuit have majority membership on each of these bodies and scientists must be stationed in the Mackenzie Delta, reproducing the kinds of long-term and place-based relationships that activist-scientists encouraged in the 1960s. In addition, the governmental appointees—whether territorial or federal—represent themselves, and not the public interest. Since their founding, Inuvialuit comanagement agencies have screened and evaluated more than seven hundred proposals, and no development has been licensed without Inuvialuit approval. Inupiat in Alaska quickly recognized the success of these institutional and constitutional arrangements. In 1983, the Inuit Circumpolar Council commissioned Thomas Berger to survey Alaska Natives' responses to the Alaska Native Claims Settlement Act, especially the act's failure to establish measures for Inuit to control decisions affecting their lives. Berger's 1985 book, *Village Journey*, argued persuasively for self-government in Alaska, to build upon the mechanisms realized through the Inuvialuit Final Agreement in Canada.[57]

As Inupiat and Inuvialuit earned say in environmental planning processes, they desired more control over the movement of scientists in general—

to be sure all forms of research respected Aboriginal interests. In 1980, as the US Navy phased out its management of the Naval Arctic Research Laboratory in Barrow, Alaska, the North Slope Borough saw an opportunity. It became a leading partner in the laboratory's operations. The borough also created a $10,000 "Arctic Science Prize," first awarded in 1984, to draw the attention of natural scientists to Inupiat needs—and recognize excellence in community-oriented, region-specific scholarship. Meanwhile, as Inuvialuit hammered out the details of their land claim with the Canadian federal government, they advocated for an updated research-licensing program within the territorial government of the Northwest Territories. In 1974, the government of the Northwest Territories amended the Scientists' and Explorer's Licensing Ordinance, which required all scientists seeking to perform fieldwork in the Northwest Territories to consult with the communities most likely to be affected by their research. In the first decade of the amended ordinance, Inuit communities made recommendations to scientists to adjust their agendas to incorporate local concerns and even denied licenses until such adjustments were made. Some scientists complained. In some cases, researchers had to present plans before receiving full permission, and scientists considered this costly and time intensive. Beyond confirming the teeth of research licensing protocols, scientists' grumbles evince their continued growing pains after land claims agreements. Scientists had been displaced from centralized decision-making structures in favor of more participatory and democratic ones.[58]

High-profile reviews of Arctic pipelines and the settlement of Inuit land claims in the 1970s thus catalyzed dramatic changes across North America, in both intended and unintended ways. Especially in the Canadian mind, Thomas Berger's final report to the National Energy Board, *Northern Frontier, Northern Homeland*, has crystallized these changes and their driving forces. In the title, Berger implied that the tensions between those supporting Arctic oil development, and those against it, stemmed from the incommensurability of two visions of the north held by southerners and indigenous peoples. Such conceptions have been oft repeated over the last forty years. In that time, Inuit, political scientists, scholars of postcolonial studies, and Arctic anthropologists have critiqued the institutions spawned by land claims agreements as being at odds with "traditional" forms of governance and knowledge making. The persistence of a scientific and political discourse in particular, as anthropologist Mark Nuttall has argued, remains a problem for Native northerners "attempting to participate in decision-making processes." While I also identify regional corporations, comanagement boards, impact assessment organizations, and research licensing protocols as sites through which the colonial encounter continues today, I am cautious about extending such an argument to the Inuvialuit Settlement Region, the North Slope Borough, or the Northwest Arctic Borough without a few caveats.[59]

Especially after reviewing a century of science, intervention, and environmental transformation in this corner of the Arctic, we can see in brighter context a set of commensurable ideas between indigenous and nonindigenous peoples, especially activist-biologists and politically organized Inuit. The postcolonial moment of the 1970s and 1980s involved the recognition—by Inuit, industry, government, and researchers—of disturbance in the Arctic, both social and ecological. In turn, a critical and humanist science of Arctic nature fostered a different set of relations within the north and between north and south. Importantly, activist-scientists and Inuit eschewed a means of organizing knowledge production that failed to include or serve northern interests. Issues of development and environmental protection, they agreed, should also be taken up together, in a regional context, through participatory decision-making processes, grounded in an appreciation of historical legacies of science and colonialism, with the best, interdisciplinary knowledge available.

Today's residents of the Beaufort Sea littoral have certainly expressed dissatisfaction with bureaucratic processes that do not sufficiently account for their voices, especially as multinational oil companies and climate scientists have descended on the region in the last decade.[60] Yet Inupiat and Inuvialuit have also professed confidence in their abilities and institutions to manage lands in a manner acceptable to beneficiaries, even with the uncertainties of global energy markets and transforming ecologies. Aboriginal groups elsewhere in North America have singled out the Inuvialuit Final Agreement—and its set of wildlife, fisheries, and development boards—as a true example of comanagement, in contradistinction to those in other modern treaties.[61] On the international level, Inupiat and Inuvialuit have also leveraged their prior experiences with land claims negotiations and pipeline proposals through the Inuit Circumpolar Council. In other words, Inuit have elaborated bureaucratic and scientific frameworks to enhance Inuit rights, environmental concerns, and interests in economic development.

I draw these conclusions not to paint a rosy picture of the 1970s and 1980s, but rather to bring nuance and complexity to portrayals of the Arctic. The structures of science and political economy in the western Arctic differ slightly across the international border—and they differ substantially from other sections of Alaska, the Canadian north, and the circumpolar world. These similarities, differences, and particularities must be appreciated in order to understand Arctic environmental change and address it fully. For these reasons, the interpretations I have offered in this book should compel today's scientists, policy makers, and historians to action. We must work against the tide of popular media that also submerges the Arctic's postcolonial nature in a sea of concern about modern climate change and globalization.

Epilogue: Unfrozen in Time

> There are two kinds of Arctic problems, the imaginary and the real. Of the
> two, the imaginary are the more real; for man finds it easier to change the face
> of nature than to change his own mind.
>
> Anthropologist-Explorer Vilhjalmur Stefansson

The signing of the Inuvialuit Final Agreement on June 5, 1984, marked the
end of a century of science, intervention, and environmental transformation
in the Arctic. The next day, another century began: the one we now inhabit.
There will be scientific attention. There will be intervention. And there will
be environmental transformation. But what kinds? The answers depend on
how we interpret the past.[1]

While scientists and the popular press have rigorously documented a
range of social and ecological conditions in the Arctic today, their narrow
take on history has created perverse effects. According to a recent survey by
the Walter and Duncan Gordon Foundation, there is widespread agreement
among northerners and southerners—those living in southern Canada and
the continental United States—that climate change is the leading issue facing the top of the world. Survey respondents identified the loss of ice, both
in permafrost and in northern seas, as indicators of an Arctic in the midst of
unfreezing. There is a startling disconnect between northerners and southerners, however, regarding what actions society should take in response.

Southerners demonstrated little awareness of Inuit sovereignty and the
channels of decision making forged in the late 1900s to negotiate environmental issues and economic development. At the expense of existing organizations, like Native governments or joint Native-federal committees, southerners promoted federal governments, through military action and improved
diplomacy among nation-states.[2] In addition to dehumanizing the north, their
proposed solutions mask metropolitan North America's responsibilities for
Arctic change—whether in the past or present—under the veil of concern

about planetary sustainability. Not coincidentally, these recommendations match those repeated in the press by economists, policy analysts, and environmental groups. The most troubling aspect of the survey results is the discernible blind spot within a citizenry compelled to do good. Despite a growing understanding of climate change, many residents of the southern United States and Canada cannot recognize the historical and moral dimensions of the issue.

To respond attentively to global change, we must not turn from history. We must learn from it. I have focused my attention on a corner of the circumpolar basin, and tracked within it the production of knowledge about nature and the changing natures of that knowledge. The resulting picture of the Arctic is by no means complete, but it is more humanizing than that of "New North" narratives. In this telling of history, the activities of scientists anchor the Arctic within a series of changes that shaped the world, from exploration and industrialization to world war to environmentalism and indigenous rights movements. Starting in the late 1800s, itinerant researchers sought out northern communities to answer questions about nature and culture and thus launch their careers as scientists. These field-workers make clear the role of Arctic landscapes in the annals of a number of disciplines, from anthropology to zoology. Yet scientists did not just seek to study northern environments. They also sought to steward them from afar. Unlike whalers, trappers, or missionaries, scientists circulated artifacts of northern life as currency in an economy of knowledge. There, the value of an item depended less on consumer choice than the administrative question du jour—border disputes at one point, defense concerns at another, and a variety of resource exploitation ventures throughout. Over the 1900s, science became so entrenched within governmental ideologies and capitalist culture in North America that its ties to interventions were rarely challenged. Many Americans and Canadians never visited the Arctic and thus relied on representations deemed scientific to legitimize governmental interference in northern affairs. Just as the development of science in the twentieth century owes debts to the Arctic, the shape of a region long considered "untouched" bears the imprint of science and colonialism.

The ethics underpinning land claims agreements and impact assessment began as undercurrents in a torrent of military-industrial activity in the Arctic, well before the recognition of global climate change. Yet they remain relevant in our time—if we can make the historical connections, that is. In the 1970s, Inuit political organizers teamed with a group of concerned researchers to advocate for indigenous sovereignty and ecological protection in the midst of plans for massive energy development. As it did then, consultative involvement continues to operate as a space of encounter between nonindigenous society and indigenous society, and the changing world we share. These procedures were born not to deny science in favor of traditional knowledge, but rather to reject the exclusionary practices of knowledge making in general. They reconfigured asymmetrical power structures, enrolling science in the service of Inuit and a public beyond a cadre of administrators. They redefined the value of knowledge as a *process* of regular exchange and open criticism, rather than a *product* of some analytical method or technology. Consultative involvement today thus constitutes both the living history of colonialism and a possible pathway toward a more equitable future on a warming planet.

What happened to these commitments to consultative involvement, to situated science, and to history? How did "New North" narratives subvert a richer understanding of change and reinstate colonial orientations to the Arctic? We find answers starting in the late 1980s, with the formation of the Arctic Council. The council is an international forum comprised of foreign affairs officials from all nation-states with Arctic territories, along with permanent participation from many of the region's indigenous peoples. The group cannot legislate, but it has shaped international and domestic affairs through reports and testimony at high-level meetings. Coming into being at the end of the Cold War, the council's mission and activities speak to that cultural moment. It promotes global stability and peace through cooperation across borders and discussion across disciplines. It seeks to connect circumpolar landscapes and communities to amplify the voices of Native northerners, pressure national governments to regulate industry, enhance human rights protections, and avoid environmental destruction. The group relies on the framework of impact assessment to gather knowledge about pressing circumpolar issues and build support for its policy proposals. Importantly, Inuvialuit and Inupiat members have guided the council's strategies over the last thirty years, even though they do not compose a majority of the indigenous participants. While the council deserves it own book-length history, the representations of the Arctic that have emerged through it since the 1980s offer a fitting ending to this one. These representations remind us that responding attentively to climate change and globalization requires constant and careful historical interpretation.[3]

THE VOICE OF THE ARCTIC COUNCIL

During the 1990s, Inuit leaders within the Arctic Council and climate scientists worked together to promote the circumpolar basin as a barometer of climate change. They became convinced that climatic changes were "more pronounced in the Arctic region," and northern ecosystems were critical to understanding "global-scale climatic processes." Accordingly, in 2000, the Arctic Council signed the Barrow Declaration, which endorsed and adopted an Arctic Climate Impact Assessment, a joint project of a few working groups within the council and the Intergovernmental Panel on Climate Change. Inuit members of the council took leading roles in the report's creation, as prior experiences with land claims agreements and environmental reviews of pipelines in North America had revealed the transformational power of science in Arctic policy. When completed in 2004, the Impact Assessment Report sprawled over eighteen chapters, documenting changes in freshwater, marine, and tundra ecosystems; shifts in hunting, herding, fishing, and gathering; and expected challenges for human health, infrastructure, and wildlife management.[4]

In their executive summary, the scientific and indigenous authors characterized the effects of climate change in terms similar to contemporary "New North" narratives, but with some glaring differences. They plainly stated the Arctic was "extremely vulnerable" to "projected climate change and its impacts." But they also underlined that climate change was one of multiple stressors in the region. It should be understood, they wrote, in light of previous land use changes, a rapid growth in the human population, and "cultural, governance, and economic" factors. Inuit representations of the Arctic implicate their own interventions, just like those of scientists in the last one hundred years. In this case, Inuit highlighted both the culpability of non-Arctic societies in northern environmental transformations and the need to recognize the layered landscapes of the circumpolar zone. The summary's conclusion is prescient, given what has happened since its release: "Impacts on the environment and society result not from climate change alone," the authors wrote, "but from the interplay of all these changes." In 2005, Inuit leaders within the Arctic Council—namely, Sheila Watt-Cloutier—pointed to the Arctic Climate Impact Assessment report to convince the United Nations to endorse global warming as a human rights issue. When the United Nations did so in 2005, it was evidence of scientists and Inuit steering international discourse on a key environmental issue—through the use of combined ecological, cultural, and historical knowledge. Scientists who contributed to the Arctic Climate Impact Assessment credit the report and Sheila Watt-Cloutier's advocacy with cementing Arctic warming as a central political issue of the modern era.[5]

Just a few years later, this view of the Arctic and its global and historical relations was sidelined. A single species became a useful prop for climate change communication, with unfortunate side effects in the social response to Arctic warming. Al Gore famously invoked the polar bear stranded on an ice floe in his 2006 documentary, *An Inconvenient Truth*, and it has been splashed across countless articles, websites, and magazine covers since. Such representations are meant to bring into bright context the rapidity of Arctic environmental change and charge fossil fuel burners elsewhere with the task of enacting personal and political action. Yet the icon of the polar bear reestablished the north as empty of people and history, clearing space for another round of intervention. In the years following the movie, US citizens supported protecting the bear as an endangered species as a reasonable climate policy. The same Inuit leaders who helped marshal the Arctic Climate Impact Assessment chafed against this misuse of the animal. At a US Senate hearing convened to consider listing the species, Inuk Mary Simon testified that safeguarding the bear would not curtail carbon emissions. Rather, it would present another challenge to indigenous residents adapting to Arctic warming by restricting hunting. For this lack of foresight in responding to climate change, Simon blamed "the media, environmental groups, and the public" for characterizing the Arctic in "overly simplistic black-and-white terms." Despite Simon's direct access to Congress, her critiques of Arctic representations were overshadowed by drama on a stage of higher profile. In 2007, the Norwegian Nobel Institute granted its Peace award to Al Gore and the Intergovernmental Panel on Climate Change, enshrining the Arctic's position within climate change discourse, but obscuring historical legacies in the north. In a bitter irony, the Nobel Institute chose not to recognize another nominee: Sheila Watt-Cloutier.[6]

In the early twenty-first century, a gulf emerged between two representations of the Arctic and their associated interventions—even though both ideas sought to recognize the significance of climate change. Al Gore's role in this gulf runs deeper than his film. As historian John English has detailed, when Gore came to the White House in 1994 as vice president, he introduced to the Arctic Council ideas and practices that contrasted sharply with the views of Inuit Circumpolar Council leaders and ministers of Canada's Department of Indian Affairs and Northern Development. Inuk Mary Simon sought to balance oil, mining, and other industrial ventures with a concern for the ecologies, human health conditions, and indigenous rights of the circumpolar north. Gore, however, emphasized conservation over development. He folded Arctic issues into a polar framework that included Antarctica, thus lending tacit support to environmental legislation that organizations like Greenpeace had suggested for the Southern Ocean—like bans on fishing and drilling. Importantly, Gore and other American officials also sought to move the Arctic Council closer to "pure research" and away from the practical interests of northern residents. These positions angered the Inuit Circumpolar Council and Canadian officials. They called out the US government for its failure to support indigenous human and economic rights, which had been made clear in the absence of self-determination provisions in the Alaska Native Claims Settlement Act of 1971. Inupiat and Inuvialuit in particular remained highly suspicious of so-called pure scientists. These researchers worked primarily at universities outside the Arctic, with datasets provided by satellite networks rather than field research. Their conclusions, Inuit and Canadian officials argued, often "romanticized the North" as a "pristine, untouched land."[7]

Statements emerging from the Arctic Council in recent years may thus seem contradictory—especially to those unfamiliar with the region's history. Inuit members have rejected some of the agendas of climate scientists, even as they relied on scientific knowledge to maneuver local ecologies, regional economies, and international agreements on global climate change. Meanwhile, a leading climate scientist declared that the human presence in the Arctic was comparable to Antarctica—negligible, and therefore easily disregarded in assessing environmental change—without a retort from the Canadian or American press.[8] At meetings of the Arctic Council in 2011, Inuit Circumpolar Council president Duane Smith announced that Inuit voices were being marginalized in recent intergovernmental discussions. He contended that the Secretariat for Inuit was underfunded and overloaded with scientific gray literature. Inuit leaders were thus incapable of keeping up with the research, offering comments and critiques on it, and sponsoring their own fieldwork. Most of this literature had been created in response to media portrayals of geopolitical conflict and polar bear populations, issues that distracted governmental entities—including indigenous bodies—from the root cause of sea ice loss: fossil fuel burning in industrial centers in the United States.[9]

While Inuit leaders have been frustrated with the Arctic Council, they have not abandoned it. In a 2014 meeting in Inuvik, Canada, the Inuit Circumpolar Council signed the Kittigaaryuit Declaration that urged the Arctic Council to build from its "demonstrated success and status as a high level forum" and trust that "issues between Arctic states, arising from developments outside of the circumpolar world, do not detract from the Council's important work." Inuit delegations before the Paris climate conference in 2015 pointed to the Arctic Council—and the scientific partnerships within it—as essential in ensuring their cultural, social, and economic health. Meanwhile, the US representatives of the council seem to have redirected energies elsewhere, even as they adopt a more active stance on Arctic climate change. At the end of 2015, the US Department of State convened GLACIER—the Conference on Global Leadership in the Arctic: Cooperation, Innovation, Engagement and Resilience—to bring together scientists, policy makers, and the foreign ministers of Arctic nations and key non-Arctic states. According to a press release, representatives of Arctic indigenous peoples were "encouraged to participate." The conference proceedings suggest, though, that Native northerners were not afforded the same influence on policy recommendations, funding decisions, or research priorities as other invited delegates. These positions of influence, by contrast, are guaranteed in the Arctic Council's rules of procedure. Inuit voices in global climate matters are being shunted by the very conversation they helped launch.[10]

This divergence between representations of climate change, and the pressures placed on the Arctic Council and Inuit Circumpolar Council within the international policy arena, continues a pattern I have outlined in this book. Granting that surface temperatures in the circumpolar world have few precedents in the scientific record, the ways scientists, environmentalists, and federal officials have addressed them remains consistent with the past. I am not implying that any one of these parties intends colonialism as an enterprise. Knowledge and power, as we have seen, are more subtle than that. I am asking for an acknowledgment from those in the southern United States and Canada that history matters. We can frame environmental transformations in the Arctic as either a total break with history or as both compounding and altering historical trajectories there. We can identify the relations between interventions and scientific practice, or we can ignore them—operating under the false pretense that scientists and scientific knowledge somehow stand apart from society. We can think of the Arctic as uniform or internally heterogeneous, as a zone unto itself or as intimately connected to the continental landmasses to its south. We can also imagine it as an intersection of all of these kinds of spaces. Our historical interpretations lead down different roads, toward different relationships among people and planet. But they always lead us.

"AN OPEN AND CRITICAL DISCUSSION"

We must choose to unfreeze the Arctic. It is not a time capsule of a premodern planet or the Tomorrowland that war-game planners, economic prognosticators, or environmental doomsayers make it out to be. It is a place, existing *in* time and *because of* time. Each of its layers was set down at a discrete moment, in response to a collision of human and more-than-human forces. This historical conception of nature abandons the grandeur of pristine wilderness, and whatever comfort or context such ideals offer for measuring change. But with hindsight provided by this transnational environmental history of science, we gain more than we lose. We acquire a critical sensitivity to our modern condition—about nature's agency, how we create scientific knowledge, and how scientific ideas mediate society's responses to environmental crises. This perspective also carries the responsibility to engage in civil society and civil discourse.

As with each of the preceding chapters, we can keep the microscope focused on scientists to confront much broader phenomena that are sure to come—like rapid warming and challenges to participatory democracy. Scientists are as interested as I am in producing public understanding about the Arctic, especially at a moment when interest in the region is swelling. In his recent book, *The Changing Arctic Environment*, oceanographer and Arctic Council scientist David Stone pinpoints the "widespread reach of many misconceptions" about the far north as an obstacle in addressing global climate change. Among other recommendations, he urges investments in Arctic education across the K–12 system as well as close partnerships between the next generation of environmental researchers and indigenous leaders. Yet he neglects the most critical issue of all: teaching scientists about their role in outside intervention and environmental transformation, whether past or present. This kind of schooling clearly influenced Stone to work so diligently for indigenous rights, environmental protection, and sustainable development in the Arctic. It should be a best practice for the Arctic scientific community.[11]

To be fair, there are many scientists who hold a deep appreciation for the histories of science and nature permeating the sites they study. Some, especially those who live full time in the north or work directly for Native organizations, clearly apply their research to public policies in ways respectful of northern communities.[12] Others, who return to field sites each summer, maintain similar attachments to people and place in the Arctic, and they pass these on to scores of graduate students.[13] Still others, who perform less regular research, are encouraged to incorporate Inuit interests and experiences by research licensing protocols that put scientists in direct contact with northerners.[14]

Yet there are many researchers who speak on Arctic matters without spending time in the region, without following licensing procedures, without completing internal ethical reviews, without abiding by regionally specific research protocols, and without incorporating human histories into their accounts of environmental change. Like Mark Serreze, whose PowerPoint slide show opened this book, their stories have offered useful insights into the nature of warming, but they are void of human experiences—especially the moral and material contours of colonialism. They concentrate on pixels without seeing the people, or the political effects of this choice. These scientists hold sway over the social response to global climate change. The ways they frame climate change drives the ways funds are made available, the ways scientists write grants, and thus the ways scientists collect data, understand environmental change, and confront society about it. Given what has transpired with the fallout from *An Inconvenient Truth* in terms of raising awareness of climate change—while creating wicked blind spots—researchers can no longer pretend science should be divorced from so-called practical interests. In choosing what to study and how to study it, they need to reckon with the long history of representations and interventions that precedes them.[15]

Simply put: all scientists should fully comply with the existing licensing and ethics review procedures provided by Aboriginal organizations connected with their proposed research sites. Such compliance entails doing more than federal Arctic research policies often require. There are models of successful partnerships among Aboriginal organizations and scientists in the Northern Contaminants Program in Canada and in the relationship between the US National Science Foundation and the Alaska Native Science Commission. There is also another opening through which scientists can shift their approach to incorporate the lessons of a transnational environmental history of science. Citing the erosion of public trust in expertise, and skepticism of figures of power, climate scientists have recently teamed with psychologists and communication experts to reorient the targets of their research. These scholars now see the relationship between science and society as crucial to building understanding about climate change as well as the trust necessary to take action on it. They focus on the development of what communication strategist Andrew Pleasant calls "an open and critical discussion" with policy makers and citizens, wherein climate impacts are scaled down from global models to the community level. This strategy dovetails nicely with the view of Arctic colonialism I have presented here, as both depict science not as value-free, or isolated from society, but rather as embedded within networks of power—but by no means inherently corrupt. These kinds of relationships seem to be unfolding as scientists partner with Inuit governments on environmental monitoring programs, either within state and territorial bodies, at the university level, or within the Arctic Council.[16] Yet scientists must take the lead in positioning the metropolitan United States and Canada as scenes of climate change as well, since these communities are most responsible for driving that process—not those in the far north.[17]

"THE POLITICAL STRENGTH OF PUBLIC OPINION"

Like climate scientists, policy makers and policy analysts concerned about Arctic futures must do more trade in histories of knowledge and nature. They need to recommit to the principles of ecological democracy that helped bring an end to a century of outside intervention in the Arctic. In recent years, the federal governments of the United States and Canada have inconsistently implemented their land claims agreements with Inuit and their assessments of oil development in the north. In the United States, the Bureau of Ocean Energy Management, Regulation, and Enforcement initially rushed analyses of environmental impact statements, justifying oil drilling in the Beaufort and Chukchi Seas "at any cost" without allowing its own scientists to complete data collection. Then in 2011, President Obama issued Executive Order 13580, which established an interagency working group to coordinate and facilitate the permitting and conduct of environmental reviews for offshore and onshore energy development in Alaska. The order charged the working group with sharing all project information across all levels of federal, state, and indigenous governments, and promoting dialogue among them. Yet it watered down the commitments to public comment periods in the National Environmental Policy Act and failed to grant membership in the working group to Alaska Native regional corporations or municipal bodies.[18]

Across the border, similar inconsistencies played out. In the summer of 2014, the National Energy Board of Canada approved seismic exploration outside of Baffin Island over the objections of Inuit communities and organizations. Okalik Eegeesiak, president of the Qikiqtani Inuit Association, noted that neither the government nor industry provided information about their activities or opportunities to comment on them. On the other hand, in 2010, the board completed a seven-year assessment of a natural gas pipeline—the Mackenzie Gas Project—a revised version of the Mackenzie Valley Pipeline project of the mid-1970s. Following several series of hearings, and with hundreds of stipulations for mitigating social and ecological impacts, the Canadian government approved the construction of a 1,100-kilometer pipeline system to connect Arctic natural resources with North American consumers.[19]

Despite the differences among these cases, there has been shared agreement among those Inuit involved on what they mean for the Arctic. That is, federal bodies—regardless of the political party in power—can short-circuit assessment and treaty implementation, either by neglecting the regional particularities of people and places, or by circumventing the processes altogether. To contest these representations and interventions, Inuit leaders have continued to engage channels of decision making they helped build in the 1970s and 1980s. Soon after the Baffin Island incident in Canada, the chair of the Inuit Circumpolar Council, Aqqaluk Lynge, demanded that industry and governments be "transparent and accountable" and follow agreed-upon codes of "free, prior, and informed consent." In 2008, the village of Point Hope, Alaska, filed a lawsuit against the US Mineral Management Service, claiming that the environmental analyses it produced ahead of a sale of leases in the Chukchi Sea were inadequate.[20] Federal courts agreed and forced the service to rewrite impact statements and develop environmental standards specific to the Arctic's marine ecosystems. When these statements and standards were in place, the village of Point Hope pulled out of the lawsuit and lent its support to offshore oil exploration.[21] In 2014, Nunavut Tunngavik, a group that oversees the Nunavut Land Claim Agreement of 1993, sued the government of Canada for not setting up an environmental monitoring program until seventeen years after the treaty was signed—though the Crown was compelled to do so by the agreement. The Nunavut Court of Appeal upheld the suit and awarded Nunavut Tunngavik $15 million.[22] Similar Inuit advocacy has countered campaigns by nongovernmental groups, too. In response to proposals from the Humane Society and Greenpeace to ban seal hunting and create economic sanctuaries in the north, Inuit have turned to social media to represent more accurately their rights to land, interests in development, and histories of exploitation.[23] Even though it is the duty of the federal governments to abide by treaty promises and assessment procedures, Inuit representatives have had to remind Canada and the United States of these responsibilities, as well as the cultural and ecological diversity of the Arctic. Inuit navigate bureaucracies; negotiate accords among Big Oil, science, and environmental groups; and decide between discussion and litigation. Some Inupiat and Inuvialuit understand these activities as continuing a long tradition of resilience.[24]

Federal governments must follow through on land claims implementation and impact assessment. And they must take seriously the historical complexities and legacies within the Arctic's varied communities. Doing so would relieve Inuit leaders from needing to keep in check decision-making structures that were themselves intended as systems of checks and balances. It would also send trickle-down effects throughout the courtrooms, boardrooms, and conference rooms in which the Arctic's future is debated. According to a

Walter and Duncan Gordon Foundation survey, Inuit residents of Alaska and Arctic Canada describe exhaustion with the number of requests to participate in land use decision making. They have stopped showing up to information sessions hosted by oil companies, public comment hearings on impact assessments, research licensing meetings, and even Arctic Council gatherings. They express bewilderment at the maze of project timelines, economic forecasts, and climate models presented to them. And, perhaps most disappointingly, they report disconnect between the needs they communicate—whether to federal officials, industry representatives, scientists, or Inuit organizations—and the actions taken by these groups. When asked for recommendations on fixing these issues, northern residents suggested greater collaboration among federal, territorial, and indigenous governments. They also demanded more nuanced conceptions of conditions on the ground in the Arctic.[25]

Commitments to fair and accurate representation—both in democracy and in visions of nature—are not a silver bullet for the complex issues of colonialism, climate change, and globalization. I emphasize them here not as a simple solution, but to show how historical interpretations of the Arctic—held by southerners—matter across a range of human relationships in and outside the north. If that notion still seems like a stretch, briefly consider industry's role in northern affairs over the last three decades. As reporters from Columbia University have shown, Exxon Mobil has twisted its portrayals of the region, depending on the audience, to ensure its maximum advantage. At private shareholder summits, scientists hired by Exxon Mobil and managers working for the company spoke plainly about climate change in the terms of a cost-benefit analysis. Some impacts, climate modelers said, would be good for business—like a longer open-water season for more detailed reconnaissance. And some would be bad—like higher sea levels, bigger waves, and slumping permafrost—which would jeopardize drilling equipment, processing plants, and pipelines. When crafting public policy positions on global warming, though, Exxon Mobil has consistently dismissed this same climate science as unreliable. At the same time, oil companies hired lobbying groups to conduct surveys of Inupiat views on development—and massaged the results to fabricate majority support for their activities in Alaska. For a while, these two-faced tactics worked to keep national citizenries and northern residents alike curiously complacent about the state of Arctic nature.[26]

Here is where we historians must acknowledge our obligations to northern affairs, too. We can no longer look away from the Arctic—that much is certain. But when we look, we must do so deliberately. We must treat our scholarship as relevant to contemporary debates. We should frame our research in response to the interests of Native northerners and develop mutually beneficial research agreements with Arctic communities.[27] Our skills do not qualify us to speak on behalf of Arctic residents, but they do enable us to shed a different light on the interconnected nature of global social and environmental transformation. We can help our peers in the sciences understand how knowledge and environments are historical processes, and thus how our disciplines and field sites are underlain by colonial experiences. We must translate these critical perspectives into action, not only in the places where we work but also in the places where we live. Our neighbors and elected officials consider wilderness designations and sanctuaries from industrial activity in the Arctic as appropriate climate strategies. They support expediting environmental reviews of energy infrastructure projects in coordinated bodies that do not allow for indigenous representation or extensive public comment. Historians can illuminate the dangers of these ideas and policies. We must. Otherwise, the complexity of human relationships with nature across the continent will be compressed, leaving their most troubling aspects unaddressed. In these respects, former Alaskan North Slope Borough mayor Eben Hopson's insights from the late 1970s resonate today. He warned that science and democracy can be warped if not "reinforced by the political strength of public opinion."[28]

CONSTANT AND CAREFUL INTERPRETATION

As I type, the media churns out stories about the Arctic. In late 2015, Big Oil pulled up its stakes in the Beaufort and Chukchi Seas—and threw into question a decade of investment in hydrocarbon development, from federal, indigenous, corporate, and environmental lobbies. The Mackenzie Gas Pipeline in the Northwest Territories, for instance, earned approval in 2010, only after thorough consultation, and only because of continued public scrutiny of the body organized to evaluate the impact assessment process, the Joint Review Panel. The pipeline has yet to be built, however. The companies involved, which include those owned by Inuit and First Nations governments, have asked for an extension of their lease until 2028. A similar slowdown transpired in Arctic Alaska in the fall of 2015. Shell Oil discontinued drilling there after spending more than $4 billion in lease purchases, scientific research, public reviews, and exploration.[29] Is this turn of events evidence of the volatility of the global economy and sea ice conditions to upset the best-laid plans of social and environmental impact assessment? Is it a crippling blow for Inuit governments dependent on oil resources for self-determination? Is it cause to celebrate the protection of a sensitive ecosystem? It would be easy for me to say that it is too early to know—that only time will tell. But time does not work like that. Humans tell time: in how we understand the past, how we bring those accounts to bear on the present, and what consequences our stories have for future possibilities.

And so, historians must pay attention to the stories as they break. For its part, the oil industry has cited a boom of hydraulic fracturing in the Bakken and Marcellus Shale gas plays, and a subsequent crash in gas prices, to explain its behavior. Greenpeace, on the other hand, has taken the opportunity to declare victory. They attributed Shell's withdrawal to their protests and have doubled-down on a permanent moratorium on oil extraction in the far north. Inuit leaders in Alaska and northwestern Canada also cast the present as a turning point. Unlike corporate executives and environmentalists, they foreground a different history—one that binds together science, resource development, and resilience. In a letter to Bart Cahir, a senior vice president for Imperial Oil, Inuvialuit leader Nellie Cournoyea highlighted the "exceptional achievement" of the formation of the Aboriginal Pipeline Group, one of the chief proponents of the Mackenzie Gas Pipeline. Three "geographically remote and culturally different" Aboriginal groups, she wrote, agreed upon "fair ownership" and then successfully negotiated with "four multi-national oil companies for a one-third ownership share," which included "Board representation and participation on all working committees." Additionally, federal legislation established a $500 million Mackenzie Gas Projects Impact Fund as well as individual Access and Benefits agreements. "These feats are unique," Cournoyea emphasized, "and were not easy to bring to conclusion." These kinds of revenues and funds are critical for indigenous support and supervision of environmental research, many Inuit point out. In turn, scientific knowledge may help Inuit transition to other forms of energy and development. Also speaking in late 2015, in the immediate aftermath of Shell Oil's termination of its Alaskan drilling program, president of the Arctic Slope Regional Corporation Rex Rock declared before the Arctic Energy Summit, "Make no mistake: we are not victims of a changing climate. Our Inupiaq culture is one of perseverance."[30] In all these instances, the rendering of the past determines the meaning of the present, even as a situation unfolds. Indeed, especially then.

Ultimately, unfreezing the Arctic means that as the circumpolar basin warms, it reveals a challenge thornier and more widespread than understanding change—building ethical and sustainable futures. In this struggle

to live rightly on Earth, science is an ally. But our guide is history.[31] Whatever unique phenomena they produce on the surface, environmental transformations have underlying connections with what has come before. Just as
there ought not be ignorance of this deeper meaning—this history—there
cannot be any conclusion to it either. There must be only the constant and
careful interpretation of places unfrozen in time.

Acknowledgments

My gratitude to Ariana Stuhl transcends the space I have for acknowledgments. In researching and writing this book, I have asked her to do many things—from giving up her job in Madison for a year in the Arctic to listening, day after day, year after year, to my latest frustrations and revelations. She has agreed to all of this and never complained about any of it. I will never forget, as long as I try, the night she found me at my computer literally pulling my hair out in despair. With grace, she delivered me from that place and restored my confidence for another day. Thank you, Ariana, for taking this journey with me, for offering your unwavering support, and for treating me with care on my best and worst days.

As a historian and scholar, my aim is to collect data as exhaustively as possible and interpret it as responsibly as possible. For the former task, I could not have traveled to archives and libraries across the United States and Canada, or spent time living in the Northwest Territories, if not for substantial financial support. I must thank the National Science Foundation Integrative Graduate Education and Research Traineeship Program (CHANGE-IGERT) at the University of Wisconsin–Madison; the Aurora Research Institute; and the American Philosophical Society, whose Franklin Research Grant allowed me to conduct fieldwork in 2014 to complete chapter five. During 2007–8 and 2010–11, when I lived in Inuvik, Canada, I served as a volunteer for the Frontiers Foundation. While working as a staff member in Inuvik's elementary and high schools, Frontiers Foundation provided me with housing and a small stipend for groceries.

The historian does not work alone in the archives, but with a team of librarians and specialists. I want to recognize Jane Linzmeyer, Jane Ann Shum, and the many Interlibrary Loan staff I never met at the University of Wisconsin–Madison. The library service coordinators in Inuvik, Northwest Territories—at the Aurora Research Institute, the Inuvialuit Cultural Resource Center, and the Inuvik Centennial Library—provided endless support during my time in town. I especially appreciate Cathy Cockney, Dave Stewart, and the Inuvialuit Cultural Resource Center for allowing me access to oral histories and other special documents that have been instrumental in this project. I also thank Rosemarie Speranza at the Alaska and Polar Collections at the University of Alaska–Fairbanks; Anastasia Tarmann at the Alaska State Library in Juneau; Arlene Schmuland at the Consortium Library for the University of Alaska–Anchorage; Robyn Dexter at the National Archives and Records Administration Anchorage location; and Jill L. Schneider at the US Geological Survey Alaska Technical Data Unit in Anchorage. Thank you to Rebecca Morin at the California Academy of Sciences; Kate Guay and Robin Weber at the Northwest Territories Archives; and Cathryn Walter, Sophie Teller, and many other document consultants at Library and Archives Canada. Finally, I appreciate the efforts of Barb Krieger, Sarah Hartwell, Eric Esau, and the unnamed student hourly at the Rauner Library at Dartmouth College for handling two large requests for copies.

On the matter of responsible interpretation, special notes are warranted. In many cases, the materials I treated were technical reports, dusty journals, or letters shared between civil servants and scientists. The authors were no longer alive. Achieving a critical distance from these documents involved little more than the typical duties of scholarly analysis—selecting a conceptual framework, performing a close reading, positioning the primary source against existing historiography, and the like. In other cases, the source was and is a breathing person—one with much at stake in how their lives and ideas are portrayed on the page. For opening themselves to me and opening my mind to alternative interpretations of the past, I am grateful to the scientists, governmental representatives, resource specialists, and residents of Inuvik, Northwest Territories, who participated in this research. In addition, as conversations with them led me to rethink Arctic history and my approach to it, I especially want to thank Duane Smith, Nellie Cournoyea, Cathy Cockney, Alana Mero, Jimmy Kalinek, Meltzer Sydney, Wayne Allen, Norm Snow, Peter Clarkson, Georgina Stefansson, Anna Fraser Pingo, Rebecca Jane Dale, Darrell Christie, Dave Button, Dick Hill, Terry Chapin, and Chris Cunada. Similarly, I want to thank the residents of Anaktuvuk Pass; the staff at Sir Alexander Mackenzie School during 2007–8, Samuel Hearne Secondary School during 2010–11, and East-Three School during 2013–15;

the students of the 2010-11 northern studies classes; and the members of the 2011 Ivvavik Field Program. While I am sure some of you may disagree with how I present Arctic history here, please know you all have my deep respect.

What appears as "my" analysis in this book thus reflects the thoughts, efforts, and experiences of many other people. Importantly, I have benefited from the feedback of several communities of scholars as well. At the University of Wisconsin, Gregg Mitman never stopped pushing me in my pursuit of this topic. He held high standards for scholarship and I am grateful he did not let me settle for anything less. I continue to glean insights from Rick Keller, Bill Cronon, Richard Staley, and Tina Loo—as well as from members of several writing groups; the Department of the History of Science; the Center for Culture, History, and Environment (CHE) of the Nelson Institute for Environmental Studies; and the CHANGE-IGERT program, especially Robert Beattie and Carmela Diosana. Anna Zeide, my ally and mentor, has been there for every milestone of this book project, offering her comforting smile and her thoughtful comments—usually on documents she has read more than once! For opening my eyes to many social, historical, and ecological issues in Alaska, I thank Terry Chapin, Vladimir Alexeev, and the participants of the 2009 summer field studies course, Global to Local Interactions: Socio-Ecological Resilience in the Rapidly Changing North, offered by the International Arctic Research Center and the Resilience and Adaptation Program at the University of Alaska-Fairbanks. I heard many helpful questions from audience members at presentations I gave at the History of Science Society and American Society for Environmental History annual meetings, the American Historical Association conferences, the Inuit Studies Conference, CHE graduate student symposia, history of science brown bags, and CHANGE lunch meetings. Thank you to Stephen Bocking, Brad Martin, and the organizers of the workshops in Whitehorse in 2009 and Peterborough in 2011. At Bucknell University, the Environmental Humanities Group helped me think through the broadest questions of the book and the argument of chapter five. Finally, I am indebted to two anonymous reviewers who provided critical readings of the book proposal and a draft of the manuscript. I hope this book reflects half of what I have learned from these reviewers and my colleagues over the last ten years.

At the University of Chicago Press, I thank Karen Merikangas Darling and Evan White for their patience and guidance in this book's journey. I also thank Dawn Hall for copyediting the entire manuscript with incredible attention to detail.

Michael Lewis, your passion for environmental history lit a fire in me. Your dedication as a mentor kept that flame going even when I was no longer your student. For that I owe you much more than I will ever be able to repay.

Dad, I wish you were here to see this.

Everett and Whitley, thank you for giving me the chance to stretch and challenge myself in ways I never knew were possible. I am very happy to have completed this project, but even happier to be growing alongside you as a father. I love you.

Archival Collections

APR-UAF   Alaska and Polar Regions Collections. Elmer F. Rasmuson Library. University of Alaska–Fairbanks.

American Association for Advancement of Science, Alaska Division, 1950–69 Records.
George L. Collins Papers.
Kathleen Lopp Smith Family Papers.
Lawrence J. Palmer Papers.
Lomen Family Papers.
Louis DeGoes Papers.
Middleton Smith Papers.
Naval Arctic Research Laboratory Records.

ARI   Aurora Research Institute Special Collections, Inuvik, Northwest Territories, Canada.
ATDU   US Geological Survey, Alaska Technical Data Unit. Anchorage, Alaska.
DBC   Dave Button Collection.

A collection of rare historical documents from the Inuvialuit Settlement Region made available by Dr. Button on his research website, Cape Krusenstern (Nuvuk), NWT (NU): Profile of an Inuit Trading Post, 1935–1947, http://www.capekrusenstern.org/research.html.

DHC   Dick Hill Collection. Inuvik Centennial Library. Inuvik, Northwest Territories.
DHPC   Dick Hill Private Collection.

Hill sent me a number of scanned documents relating to his participation in the Mackenzie Institute during the 1960s and early 1970s.

ICRC   Inuvialuit Cultural Resource Center. Inuvik, Northwest Territories. Permission to use informally published and unpublished materials.

Cathy Cockney, *Kitigaaryuit Oral Traditions Research Project 1996: English Translations and Transcriptions of Interview Tapes #1–17.*
Cathy Cockney, *Kitigaaryuit Oral Traditions Research Project 1997: English Translations and Transcriptions of Interview Tapes #1–16.*
Arnold, Charles, Wendy Stephenson, Bob Simpson, and Zoe Ho, eds. *Taimani: At That Time; Inuvialuit Timeline Visual Guide.* Inuvialuit Regional Corporation, 2011.

Elisa J. Hart, *Reindeer Days Remembered*, 2001
Elisa J. Hart and Cathy Cockney, *Yellow Beetle Oral History and Archaeology Project*, 1999.
Murielle Ida Nagy, "Aulavik Oral History Project on Banks Island, NWT: Final Report," 1999.
Murielle Ida Nagy, "Yukon North Slope Inuvialuit Oral History," 1994.
Randal Pokiak, *Inuvialuit History*, 1991.

LAC   Library and Archives Canada. Ottawa, Ontario.

Alf Erling Porsild Fonds, R5495-0-7-E.
J. J. O'Neill Fonds, MG 30 B 171.
Kenneth Gordon Chipman Fonds, MG 30 B 66.
R. F. Legget Fonds, MG 30 J 44.
Rudolph Martin Anderson Fonds, MG 30 40.
RG 33, Records of Royal Commissions.
RG 42, Department of Marine Fonds.
RG 45, Geological Survey of Canada.
RG 92, Geographical Branch.
RG 109, Canadian Wildlife Service.
RG 132, National Herbarium.
T-13267; T-13273; T-13323; T-13324; T-13329; T-14212,
Northwest Territories and Yukon Branch.

NARA-A   National Archives and Records Administration at Anchorage. Anchorage, Alaska.

RG 75, Bureau of Indian Affairs.
RG 181, Records of Naval Districts and Shore Establishments.

NARA   National Archives and Records Administration at College Park, MD.

M641. Letters Received by the Revenue-Cutter Service, August 11, 1869–September 28, 1910.
RG 22, US Fish and Wildlife Survey.
RG 57, US Geological Survey.
RG 401, Collection 46, Alton A. Lindsey Papers.
RG 401, Collection XJCR, John C. Reed Papers.
RG 401, Collection 95, Paul C. Dalyrmple Papers.

NWTA   Northwest Territories Archives. Prince of Wales Heritage Center. Yellowknife, Canada.

Canada, Advisory Commission on the Development of Government in the Northwest Territories Fonds, N-1992-221.
Canada, Interdepartmental Reindeer File, G-1979-069.
Canada, Northern Administration Branch, G-1979-003.
Canadian Broadcasting Company Transcripts, N-1992-014.
"C. T. Pedersen," Public reference file.
"Distant Early Warning Line," Public reference file.
"Inuvik," Public reference file.
Journal of Frank Russell, N-2002-037.
Northern Newsletter Collection, N-1998-037.
"Reindeer," Public reference file.
Sven Johansson, "The Canadian Reindeer Herd," N90-002.
Transcript of Interview with G. W. Porter, N-1992-173.
"Whaling," Public reference file.

PEKL   Papers of Ernest de Koven Leffingwell, 1900–1961. Rauner Special Collections. Dartmouth College Library. New Haven, Connecticut.

PSC   Papers of Vilhjalmur Stefansson, 1902–62. Rauner Special Collections. Dartmouth College Library. New Haven, Connecticut.

UAA   University of Alaska–Anchorage. Consortium Library. Archives and Special Collections.

David M. Hickok Papers.

Dorothy and Grenold Collins Collection

Notes

INTRODUCTION

1. Mark Serreze, "The Arctic as the Messenger of Global Climate Change," presentation to the Inuit Studies Conference, Washington, DC, October 26, 2012.

2. For this sample of "New North" narratives, see A. L. Parlow, "Shell and Beyond: Toward an Arctic Standard in the New North," *Alaska Dispatch*, July 29, 2012; "The New North," *Nature* 478 (October 13, 2011): 172–74. Gwynne Dyer, *Climate Wars: The Fight for Survival as the World Overheats* (Oneworld, 2011); Bob Reiss, *The Eskimo and the Oil Man: The Battle at the Top of the World for America's Future* (New York: Business Plus, 2012); and Greenpeace, "Save the Arctic," http://www .greenpeace.org/usa/arctic/, accessed October 15, 2015.

3. The literature on polar exploration is immense, and cannot possibly be summarized here. For a primer on connections between geographical knowledge and imperial power in the search for the North Pole, see Neil Smith, *American Empire: Roosevelt's Geographer and the Prelude to Globalization* (Berkeley: University of California Press, 2003), 83–112; Lisa Bloom, *Gender on Ice: American Ideologies of Polar Expeditions* (Minneapolis: University of Minnesota Press, 1993), 2; and Trevor H. Levere, *Science and the Canadian Arctic: A Century of Exploration, 1818-1918* (New York: Cambridge University Press, 2004), 338–76.

4. The best source for understanding the relations between science and culture in polar exploration in the United States is Michael F. Robinson, *The Coldest Crucible: Arctic Exploration and American Culture* (Chicago: University of Chicago Press), 2006. On divergent Inuit representations, see Ann Fienup-Riordan, *Freeze Frame: Alaska Eskimos in the Movies* (Seattle: University of Washington Press, 2003). For the imperial visions of scientific travel in general, see Mary Louise Pratt, *Imperial Eyes: Travel Writing and Transculturation* (New York: Routledge, 1992).

5. Ian Hacking, *Representing and Intervening: Introductory Topics in the Philosophy of Natural Science* (Cambridge: Cambridge University Press, 1983), 130–31. Helen Tilley's analysis of environment, knowledge, and empire employs Hacking's framework and thus serves as a model for my approach to the Arctic. See Helen Tilley, *Africa as a Living Laboratory: Empire, Development, and the Problem of Scientific Knowledge, 1870-1950* (Chicago: University of Chicago Press, 2011).

6. Historians have treated in depth the cultural history of northern representations, though they have not always considered the material effects of these ideas in their analysis. Sherrill E. Grace, *Canada and the Idea of North* (Montreal: McGill-Queen's University Press, 2001); and Susan Kollin, *Nature's State: Imagining Alaska as the Last Frontier* (Chapel Hill, NC: University of North Carolina Press, 2001).

7. While Gwich'in residents have occupied and used parts of this selected transnational region, I do not focus on their interactions with scientists as much as I do Inuit. Other historians have

deconstructed the politics and history of Arctic science—but much of their work has treated the region as a whole, or certain parts of it, namely, Russia and the "Nordic" regions of Scandinavia and Greenland. See Sverker Sörlin, *Science, Geopolitics, and Culture in the Polar Region: Norden beyond Borders* (Burlington, VT: Ashgate 2013); Michael Bravo and Sverker Sörlin, eds., *Narrating the Arctic: A Cultural History of Nordic Scientific Practices* (Canton, MA: Science History Publications, 2002); John McCannon, *A History of the Arctic: Nature, Exploration, and Exploitation* (London: Reaktion Books, 2012); and Paul R. Josephson, *The Conquest of the Russian Arctic* (Cambridge, MA: Harvard University Press, 2014). Scholars who do not identify as academic historians have challenged ahistorical conceptions of the Arctic and brought attention to the historical legacies of colonialism present on the contemporary northern scene. See Emilie Cameron, *Far Off Metal River: Inuit Lands, Settler Stories, and the Making of the Contemporary Arctic* (Vancouver: University of British Columbia Press, 2015); Robert McGhee, *The Last Imaginary Place: A Human History of the Arctic World* (Chicago: University of Chicago Press, 2005); and Mark Nuttall, "Epistemological Conflicts and Cooperation in the Circumpolar North," in *Globalization and the Circumpolar North*, ed. Lassi Heininen and Chris Southcott (Fairbanks: University of Alaska Press, 2010), 149–78. These authors do not consider the range of scientists or the particular landscape I examine here.

8. The best example of this Alaskan grand narrative can be found in Stephen Haycox, *Frigid Embrace: Politics, Economics, and Environment in Alaska* (Corvallis: Oregon State University Press, 2002). Haycox argues that Alaska's American history can be understood as the gradual shift from a place of natural wealth converted into materials of capitalist opportunity to a wilderness protected from economic development. See also Claus-M. Naske and Herman E. Slotnick, *Alaska: A History* (Norman: University of Oklahoma Press, 2011); and Roxanne Willis, *Alaska's Place in the West: From Last Frontier to the Last Great Wilderness* (Lawrence: University Press of Kansas, 2011). On the Canadian side, see Kenneth Coates, *Canada's Colonies: A History of the Yukon and Northwest Territories* (Toronto: James Lorimer, 1985). Coates believes it is fair to refer to Yukon and Northwest Territories as "colonies" because of the physical and psychological distance between the land and people and the forces of government that imposed order on the region. See also Morris Zaslow, *The Opening of the Canadian North, 1870–1914* (Toronto: McClelland and Stewart, 1971); Arthur Ray, "Recent Trends in Northern Historiography," *Essays on Canadian Writing* 59 (Fall 1996); Kenneth S. Coates and William R. Morrison, "The New North in Canadian History and Historiography," *History Compass* 6, no. 2 (2008): 639–58; Ken S. Coates, ed., *Arctic Front: Defending Canada in the Far North* (Toronto: Thomas Allen, 2008); Francis Abele, "The State and the Northern Social Economy: Research Prospects," *Northern Review* 30 (Spring 2009): 37–58; Donald Worster addresses the differences between the American "Turnerian" model and the Canadian "metropolitan" model of national development in Donald Worster, "Two Faces West: The Development Myth in Canada and the United States," in *Terra Pacifica: People and Place in the Northwest States and Western Canada*, ed. Paul W. Hirt (Pullman: Washington State University Press, 1998), 71–87.

9. In addition to those listed above, I am indebted to scholars of the Canadian north who directly treat scientists as actors in colonial history. There has been some excellent work on Arctic science in North America, though nearly all of it addresses the period after World War II. For examples, see Matthew Farish and P. Whitney Lackenbauer, "High Modernism in the Arctic: Planning Frobisher Bay and Inuvik," *Journal of Historical Geography* 35 (2009): 517–44; Stephen Bocking, "A Disciplined Geography: Aviation, Science, and the Cold War in Northern Canada, 1945–1960," *Technology and Culture* 50, no. 2 (2009): 265–90; Stephen Bocking, "Science and Spaces in the Northern Environment," *Environmental History* 12 (October 2007): 867–94; and Peter Kulchyski and Frank Tester, *Kiumajut (Talking Back): Game Management and Inuit Rights, 1900–1970* (Vancouver: University of British Columbia Press, 2007).

10. Some scholars of Alaska and the Canadian north have paved the way here, linking episodes in North America to colonial networks in Africa, and the scholarly literature on empire and environment more generally. See Cameron, *Far Off Metal River*; Robert Bruce Campbell, *In Darkest*

Alaska: Travels and Empire along the Inside Passage (Philadelphia: University of Pennsylvania Press, 2007); John Sandlos, *Hunters at the Margin: Native People and Wildlife Conservation in the Northwest Territories* (Vancouver: University of British Columbia Press, 2007); and Liza Piper, *The Industrial Transformation of Subarctic Canada* (Vancouver: University of British Columbia Press, 2009).

11. Among environmental historians working in North America, the turn toward transnational studies of science is one way to transcend nationalist frontier traditions and engender richer understandings of place. See Paul Sutter, "What Can U.S. Environmental Historians Learn from Non-U.S. Environmental Historiography?" *Environmental History* 8, no. 1 (2003): 109-29. The utility of a historical analysis of the Arctic attentive to networks—especially those of scientists—is suggested in Finn Arne Jørgensen, "The Networked North: Thinking about the Past, Present, and Future of Environmental Histories of the North," in *Northscapes: History, Technology, and the Making of Northern Environments*, ed. Dolly Jørgensen and Sverker Sörlin (Vancouver: University of British Columbia Press, 2013), 268-279. See also Ronald E. Doel, Urban Wråkberg, and Suzanne Zeller, "Science, Environment, and the New Arctic," *Journal of Historical Geography* 44 (2014): 2-14.

12. For an argument for the materialist history of science and environment I subscribe to here, see Gregg Mitman, "Living in a Material World," *Journal of American History* 100, no. 1 (2013): 128-30. Sverker Sörlin has also recently proposed this kind of analysis of the cryosphere. Sverker Sörlin, "Cryo-History: Narratives of Ice and the Emerging Arctic Humanities," in *The New Arctic*, ed. Birgitta Evengård, Joan Nymand Larsen, and Øyvind Paasche (Cham, Switzerland: Springer, 2015), 327-39. Stephen Bocking has called for environmental histories of Arctic science specifically. Stephen Bocking, "Situated yet Mobile: Examining the Environmental History of Arctic Ecological Science," in *New Natures: Joining Environmental History with Science and Technology Studies*, ed. Dolly Jørgensen, Finn Arne Jørgensen, and Sara B. Pritchard (Pittsburgh: University of Pittsburgh Press, 2013), 164-78.

13. Nellie Cournoyea, "Adaptation and Resilience: The Inuvialuit Story," presentation to the Inuit Studies Conference, Washington, DC, October 27, 2012.

14. Historians of science and empire have been especially attentive to the ways indigenous actors have shaped the production of knowledge about nature. See David Turnbull, "Boundary-Crossings, Cultural Encounters, and Knowledge Spaces in Australia," in *The Brokered World*, ed. Simon Schaffer et al. (Sagamore Beach, MA: Science History Publications, 2009), 387-428. These same scholars also use analytical approaches similar to transnational environmental history to account for the flows of knowledge, people, and material goods in and through colonial spaces. David Wade Chambers and Richard Gillespie, "Locality in the History of Science: Colonial Science, Technoscience, and Indigenous Knowledge," in *Nature and Empire: Science and the Colonial Enterprise*, ed. Roy M. MacLeod (Chicago: University of Chicago Press, 2000), 221-40.

15. Research from archaeologists and anthropologists fill in some of the gaps but cannot substitute for Inuit accounts of northern pasts. I refer to some of this scholarship, especially in analyzing the early and mid-nineteenth century, when the archival record from scientists and Inuit is thinner than in later periods. See McGhee, *Last Imaginary Place*; and Donald S. Johnson, "Northern Periphery: Long-Term Inuit-European and Euroamerican Intersocietal Interaction in the Central Canadian Arctic" (master's thesis, McGill University, 1999). Critical archaeologists have also collaborated with Inuvialuit to renarrate history. Natasha Lyons, "*Quliaq tohongniaq tuunga* (Making Histories): Towards a Critical Inuvialuit Archaeology in the Canadian Western Arctic" (PhD diss., University of Calgary, 2007), especially 75-88.

16. Donald Worster, "Doing Environmental History," in *The Ends of the Earth: Perspectives on Modern Environmental History*, ed. Donald Worster (New York: Cambridge University Press, 1989), 289-308. Neil Safier, "Global Knowledge on the Move: Itineraries, Amerindian Narratives, and Deep Histories of Science," *Isis* 101 (2010): 133-44. On public, engaged humanities, see Gregory Jay, "What (Public) Good Are the (Engaged) Humanities?" *Imagining America*, http://imagining america.org/fg-item/what-public-good-are-the-engaged-humanities/, accessed December 2, 2015.

On "close following" rather than "participant observation," see Katharine Cramer, "Scholars as Citizens: Studying Public Opinion through Ethnography," in *Political Ethnography*, ed. Ed Schatz (Chicago: University of Chicago Press, 2009). "Inuvialuit Regional Corporation Guidelines for Research in the Inuvialuit Settlement Region," http://nwtresearch.com/sites/default/files/inuvialuit -regional-corporation.pdf, accessed December 2, 2015.

17. Emilie Cameron eloquently summarizes the responsibilities of academic scholars in the context of postcolonial, community-engaged research about Arctic places and indigenous histories. See Cameron, *Far Off Metal River*, 20–29. Following her model, I share some more details on my research practice. This research followed the licensing protocols of the Aurora Research Institute in Inuvik, Northwest Territories. After receiving a license for research, I followed the feedback of the community representatives who reviewed my proposal—which requested that I submit a copy of my study to the town of Inuvik when complete. I delivered a bound copy of my dissertation to the library at the Aurora Research Institute and an electronic copy to the CEO Inuvialuit Regional Corporation in May 2014. I sent a draft of the manuscript to several residents of Inuvik who advised me during the research. See also Ned Searles, "Why Do You Ask So Many Questions? Learning How Not to Ask in Canadian Inuit Society," *Journal for the Anthropological Study of Human Movement* 11, no. 1 (2000): 247–64; and Linda Tuhiwai Smith, *Decolonizing Methodologies: Research and Indigenous Peoples* (New York: Zed Books, 1999), especially 20–43.

18. For examples, see Abraham Okpik, "What Do the Eskimo People Want?" *Northern Affairs Bulletin* 7, no. 2 (1960): 38–42; Abraham Okpik, *We Call It Survival: The Life Story of Abraham Okpik*, ed. Louis McComber, Life Stories of Northern Leaders 1 (Iqaluit, Nunavut: Nunavut Arctic College, 2005); Nuligak, *I, Nuligak*, ed. and trans. Maurice Metayer (Toronto: Peter Martin, 1966); and Alice French, *My Name Is Masak* (Winnipeg: Peguis Publishers, 1976).

19. My commitment to a place-based Arctic history that spans the nineteenth and twentieth centuries, and pays attention to ecology, anthropology, and indigenous studies, follows the work of Lyle Dick. Lyle Dick, *Muskox Land: Ellesmere Island in the Age of Contact* (Calgary: University of Calgary Press, 2001).

20. On history as constitutive of the present, see Dipesh Chakrabarty, *Provincializing Europe* (Princeton, NJ: Princeton University Press, 2007), 1–22.

21. One trend within the literature on colonialism in India and Africa is to narrate indigenous independence movements as involving the erosion of science. See Gyan Prakash, *Another Reason: Science and the Imagination of Modern India* (Princeton, NJ: Princeton University Press, 1999); and Julie Livingston, *Debility and the Moral Imagination in Botswana* (Bloomington: Indiana University Press, 2005). Upon recognition of their land claims, Inuit in Alaska and western Canada have invested in language programs, cultural revitalization, and the promotion of "local ecological knowledge." These events have been the subject of scholarly research, especially from anthropologists. I am interested here in how Inupiat and Inuvialuit have turned squarely to science as a means of self-determination, particularly in matters of land use, development, and environmental management. I also want to emphasize that Inuvialuit and Inupiat identity, and Arctic life more generally, is not only a response to colonialism and its legacies. I choose to emphasize colonialism in these pages because it is so often unmentioned in the dominant narratives about the Arctic today.

22. "Presentism" remains a genuine concern for historians, though the concept is often misunderstood. Lynn Hunt, "Against Presentism," *Perspectives on History* (May 2002), https://www .historians.org/publications-and-directories/perspectives-on-history/may-2002/against-presentism, accessed November 20, 2015.

23. Environmental historians have been uniquely concerned with speaking to natural and social scientists about the relevance of history to environmental crises. See R. Hoffman, N. Langston, J. McCann, P. Perdue, and L. Sedrez, "AHR Conversation: Environmental Historians and Environmental Crises," *American Historical Review* 113 (2008): 1431–65.

24. Emilie S. Cameron, "Securing Indigenous Politics: A Critique of the Vulnerability and Adaptation Approach to the Human Dimensions of Climate Change in the Canadian Arctic," *Global*

Environmental Change 22 (2012): 103–14. Terry Fenge and Paul Quassa, "Negotiating and Implementing the Nunavut Land Claims Agreement," *Options Politiques* (July–August 2009): 80–86.

25. Sheila Watt-Cloutier, *The Right to Be Cold: One Woman's Story of Protecting Her Culture, the Arctic, and the Whole Planet* (Toronto: Allen Lane, 2015).

26. Leah Glaser, "Public Historians Take on Climate Change," *Public History Commons*, http://publichistorycommons.org/public-historians-take-on-climate-change/, accessed December 1, 2015. This article summarizes the presentations by Carey and Langston at the "Historians and Climate Change Panel" before the National Council on Public History and Organization of American Historians conference in Milwaukee during April 2012.

27. The historical literature on these "last frontiers" is more developed than that of the North American Arctic, and becomes sources for comparison in many of the chapters that follow. See, for example, Sverker Sörlin, *Science, Geopolitics, and Culture in the Polar Region* (Burlington, VT: Ashgate, 2013); and Birgitta Evengård, Joan Nymand Larsen, and Øyvind Paasche, eds., *The New Arctic* (Cham, Switzerland: Springer, 2015). On Antarctica, see Stephen Pyne, "The Berg," and "Heart of Whiteness: The Literature and Art of Antarctica," in *The Ice: A Journey to Antarctica* (Iowa City: University of Iowa Press, 1986), 1–21; 150–207.

28. Howard Zinn, *A People's History of the United States: 1492–Present* (New York: HarperCollins, 2003), 9.

CHAPTER 1

1. Robert McGhee, *Beluga Hunters: An Archaeological Reconstruction of the History and Culture of the Mackenzie Delta Kittegaryumiut* (Saint Johns: Institute of Social and Economic Research, Memorial University of Newfoundland, 1974), 1–6. See also The Canadian Encyclopedia, "Kitigaaryuit," http://www.thecanadianencyclopedia.ca/en/article/kitigaaryuit-kittigazuit/, accessed June 18, 2015. Alternate spellings of the settlement include Kittigaaryuit and Kittigazuit.

2. This description is drawn from John Bockstoce, *Furs and Frontiers in the Far North: The Contest among Native and Foreign Nations for the Bering Strait Fur Trade* (New Haven, CT: Yale University Press, 2009), 275–89.

3. McGhee, *Beluga Hunters*, 1–6.

4. Charles Arnold, Wendy Stephenson, Bob Simpson, and Zoe Ho, eds., *Taimani: At That Time; Inuvialuit Timeline Visual Guide* (Inuvik, NT: Inuvialuit Regional Corporation, 2011), 54. During my visits to the Inuvialuit Cultural Resource Center between 2007 and 2014, the central exhibit on display concerned whaling, including but not limited to, the arrival of commercial whalers in the late 1800s. There were no displays on Royal Naval officers or the search for the North Pole. The kind of reconstruction of a period of settlement I perform here—with attention to science, commerce, and space—draws from models in postcolonial literature, particularly those by historical geographers. See Daniel W. Clayton, *Islands of Truth: The Imperial Fashioning of Vancouver Island* (Vancouver: University of British Columbia Press, 2000), and R. Cole Harris, *The Resettlement of British Columbia: Essays on Colonialism and Geographical Change* (Vancouver: University of British Columbia Press, 1997).

5. Felipe Fernández-Armesto, foreword to *Furs and Frontiers in the Far North: The Contest among Native and Foreign Nations for the Bering Strait Fur Trade*, by John R. Bockstoce (New Haven, CT: Yale University Press, 2009), xiii–xiv.

6. Helen M. Buss, *Undelivered Letters to Hudson's Bay Company Men on the Northwest Coast* (Vancouver: University of British Columbia Press, 2003), 3–12. See also John K. Stager, "Fort Anderson: The First Post for Trade in the Western Arctic," *Geographical Bulletin* 9, no. 1 (1967): 46–48. On trade relations among Inuit and Dene, see McGhee, *Beluga Hunters*, 3.

7. On MacFarlane's early career, see Debra Lindsay, *Science in the Subarctic: Trappers, Traders, and the Smithsonian Institution* (Washington, DC: Smithsonian Institution Press, 1993), 59–61.

On MacFarlane's background, see Inuvialuit Cultural Resource Center, "Inuvialuit Pitqusiit Inuu-niarutait (Inuvialuit Living History): Roderick MacFarlane," http://www.inuvialuitlivinghistory.ca /wiki_pages/Roderick%20MacFarlane, accessed July 13, 2012.

8. Natasha Lyons, Kate Hennessy, Charles Arnold, and Mervin Joe, "The Inuvialuit Smithsonian Project: Winter 2009-Spring 2011," report produced for the Smithsonian Institution, vol. 1 (June 2011), 2. See also Stager, "Fort Anderson," 54-56.

9. Lindsay, *Science in the Subarctic*, 59. See also Lyons, Hennessy, Arnold, and Joe, "Inuvialuit Smithsonian Project: Winter 2009-Spring 2011," 17. By 1867, MacFarlane had collected nearly five thousand natural history specimens on his travels between Fort Good Hope and Fort Anderson and purchased three hundred artifacts from Inuit. For more detailed descriptions of the speci-mens sent and a dynamic interface with which to explore them, visit Inuvialuit Cultural Resource Center, "Inuvialuit Pitqusiit Inuuniarutait (Inuvialuit Living History): Welcome to the MacFarlane Collection," http://www.inuvialuitlivinghistory.ca/collection, accessed July 13, 2012.

10. Daniel Goldstein, "'Yours for Science': The Smithsonian Institution's Correspondents and the Shape of Scientific Community in Nineteenth-Century America," *Isis* 85, no. 4 (1994): 573-99. The details of how Baird and MacFarlane enrolled trappers in the trade for science are presented in Lindsay, *Science in the Subarctic*, 64-70. For the use of directions on collecting and preserving so as to improve the order of collections, see ibid., 28-40. On Gibbs's instructions, see ibid., 78-88. On the paths specimens took out of northern Canada between 1859 and 1867, see ibid., 122.

11. Lyons, Hennessy, Arnold, and Joe, "Inuvialuit Smithsonian Project: Winter 2009-Spring 2011," 2-17.

12. Roderick MacFarlane, "On an Expedition down the Begh-ula or Anderson River," *Canadian Record of Science* 4 (1891): 28-53. See also E. O. Hohn, "Roderick MacFarlane of Anderson River and Fort," *Beaver* (1963): 22-26.

13. Stager, "Fort Anderson," 45-46. Emilie Cameron, "Copper Stories: Imaginative Geographies and Material Orderings of the Central Canadian Arctic," in *Rethinking the Great White North: Race, Nature, and the Historical Geographies of Whiteness in Canada*, ed. A. Baldwin, L. Cameron, and A. Kobayashi (Vancouver: University of British Columbia Press, 2011), 169-90. See also Doug Owram, *Promise of Eden: The Canadian Expansionist Movement and the Idea of the West* (Toronto: University of Toronto Press, 1992), especially chapters 1-3.

14. Quoted in Hohn, "Roderick MacFarlane of Anderson River and Fort," 24.

15. Quoted in ibid., 28.

16. James C. Scott, *Seeing Like a State: How Certain Schemes to Improve the Human Condition Have Failed* (New Haven, CT: Yale University Press, 1999), 16-17. The hope and despair circling notions of the Northwest as both a wilderness and a resource treasure also characterized Canadian and British visions of the western agricultural frontier at the same time. See Owram, *Promise of Eden*, 59-78.

17. Bockstoce, *Furs and Frontiers in the Far North*, 282-83.

18. Ibid., 275-85. See also Murielle Ida Nagy, "Yukon North Slope Inuvialuit Oral History," *Occasional Papers in Yukon History No. 1*, Heritage Branch, Government of the Yukon, 1994. Stager, "Fort Anderson," 46-48.

19. Bockstoce, *Furs and Frontiers in the Far North*, 299-302. See also Ted C. Hinckley, *The Amer-icanization of Alaska, 1867-1897* (Palo Alto, CA: Pacific Books, 1972), 20-23.

20. On Alaska's period of military rule, see Claus-M. Naske and Herman E Slotnick, *Alaska: A History* (Norman: University of Oklahoma Press, 2011), 99-122. On the support of the whaling industry, see John R. Bockstoce, *Whales, Ice, and Men: The History of Whaling in the Western Arctic* (Seattle: University of Washington Press, 1986), 290-323.

21. Bockstoce summarizes the changes to the whaling industry and its expansion up the north-west coast in Bockstoce, *Furs and Frontiers in the Far North*, 275-315. Nye is quoted on page 280. On the ecology of the Mackenzie River and Beaufort Sea, see Bockstoce, *Whales, Ice, and Men*, 263. For estimates of the bowhead whale population in the Beaufort Sea prior to the American whaling industry, see John R. Bockstoce and Daniel B. Botkin, "The Historical Status and Reduction of the

Western Arctic Bowhead Whale (*Balaena mysticetus*) Population by the Pelagic Whaling Industry, 1848-1914," *Scientific Reports of the International Whaling Commission*, Special Issue, no. 5 (1983): 107-41.

22. Eric Jay Dolin, *Leviathan: The History of Whaling in America* (New York: W. W. Norton, 2007), 120, 140, and 356. See also Bockstoce, *Whales, Ice, and Men*, 205-324.

23. Bockstoce, *Furs and Frontiers in the Far North*, 278. On the loss of whaling ships, see Hinckley, *Americanization of Alaska*, 79. On the Beaufort Sea as "Forbidden Sea," see Prince of Wales Heritage Center, "Historical Timeline of the Northwest Territories," www.pwnhc.ca/timeline, accessed July 12, 2012.

24. "Memo to the Honorable Secretary of the Treasury, Washington, DC, November 6th, 1884," in M641, Letters Received by the Revenue-Cutter Service, August 11, 1869-September 28, 1910, NARA. On the Revenue Cutter Service's work in establishing Point Barrow as a safe station, see Charles D. Brower, *King of the Arctic: A Lifetime of Adventure in the Far North* (London: Robert Hale, 1958), 12-13. A summary of the establishment of the International Polar Year Expedition can be found in John Murdoch, *Ethnological Results of the Point Barrow Expedition*, introduction by William Fitzhugh (Washington, DC: Smithsonian Institution Press, 1998), ix-xvi.

25. Murdoch, *Ethnological Results of the Point Barrow Expedition*, xii-xvi. See also Cornelia Luedecke, "The First International Polar Year (1882-1883): A Big Science Experiment with Small Science Equipment," *Proceedings of the International Commission on History of Meteorology* 1, no. 1 (2004): 56. Whaling also provided the pretext for various European nation-states to deploy science in imperial activities in the Antarctic. See Peder Roberts, *The European Antarctic: Science and Strategy in Scandinavia and the British Empire* (New York: Palgrave Macmillan, 2011), especially chapters 1-3.

26. Murdoch, *Ethnological Results of the Point Barrow Expedition*, xiv-xxxviii.

27. Hartson Bodfish, *Chasing the Bowhead* (Cambridge, MA: Harvard University Press, 1936), 41. Arthur James Allen, *A Whaler and Trader in the Arctic, 1895 to 1944: My Life with the Bowhead* (Anchorage: Alaska Northwest Publishing, 1978), 144.

28. Murdoch, *Ethnological Results of the Point Barrow Expedition*, xix. Middleton Smith, "Journal: August 12, 1882 to March 3, 1883," International Polar Year Expedition to Pt. Barrow, Middleton Smith Papers (Subgroup 1), Box 1, folder 2, APR-UAF.

29. On the construction of a station by the Pacific Steam Whaling Company, see Bockstoce, *Whales, Ice, and Men*, 317-18. On the US government's relief station, see "Appropriations, 1889-1895," M641 Letters received by the Revenue-Cutter Service, August 11, 1869 to September 28, 1910, Letters Received concerning Administration of the Point Barrow Refuge Station, 1884-1898, NARA.

30. Dolin, *Leviathan*, 336-41, 356.

31. Bockstoce, *Furs and Frontiers in the Far North*, 220-30, 350.

32. On shore whaling, see Bockstoce, *Whales, Ice, and Men*, 231-54. On the trade relations that marked the expansion of whaling in the 1880s and 1890s, see ibid., 201-2. Along the passage from the west coast of Alaska to the Beaufort Sea, whalers often gave these Inuit new English handles (like Cockney, Big Jim, Sam Brown, for instance) because they could not pronounce their Native names.

33. Two of the largest whaling settlements, Point Hope and Point Barrow, were sited at the two largest Inuit villages on the coast, Tigara and Utkiavik. See Brower, *King of the Arctic*, 12-78. On the declining importance of historical trading places, see Bockstoce, *Furs and Frontiers in the Far North*, 335-55.

34. See Bockstoce, *Whales, Ice, and Men*, 255-88. See also Nagy, "Yukon North Slope Inuvialuit Oral History," 33.

35. Bockstoce and Botkin, "Historical Status and Reduction of the Western Arctic Bowhead Whale (*Balaena mysticetus*) Population by the Pelagic Whaling Industry," 107-41. See also Rob Ingram and Helene Dobrowolsky, *Waves upon the Shore: A Historical Profile of Herschel Island*, prepared for Heritage Branch, Department of Tourism, Government of Yukon Territory, September

1989, 78. On the decline of marine mammals in the Bering Strait and the incentive to move north to trade for seal and whale, see Bockstoce, *Furs and Frontiers in the Far North*, 319–24. On disease impacts, see McGhee, *Beluga Hunters*, 1–6. For population estimates, see Froelich Rainey, "Native Economy and Survival in Arctic Alaska," *Applied Anthropology* 1 (October–December 1941): 10.

36. Ingram and Dobrowolsky, *Waves upon the Shore*, 78. On Firth, see Bockstoce, *Furs and Frontiers in the Far North*, 351.

37. Murdoch, *Ethnological Results of the Point Barrow Expedition*, xix, li.

38. Ibid., xix. Middleton Smith, "Journal: August 12, 1882 to March 3, 1883," International Polar Year Expedition to Pt. Barrow, Middleton Smith Papers (Subgroup 1), Box 1, folder 2, APR-UAF. See especially journal entry for January 20, 1883.

39. Allen, *A Whaler and Trader in the Arctic*, 65–70; 151–53. Bodfish, *Chasing the Bowhead*, 57.

40. Frank Russell, *Explorations in the Far North: Being the Report of an Expedition under the Auspices of the University of Iowa during the Years 1892, '93, and '94* (Iowa City: University of Iowa, 1898), 144–52. See also Vilhjalmur Stefansson, "Eskimo Trade Jargon of Herschel Island," *American Anthropologist* 11, no. 2 (1909): 217–32.

41. Carol Harker, "Northward, Ho!" *Iowa Alumni Magazine* (March 1992), http://www .iowalum.com/magazine/print.cfm?target_url=http://www.iowalum.com/magazine/mar92 /northward.cfm?page=print, accessed October 1, 2011.

42. On the development of universities in the late 1800s and their role in the rise of professional science, see Ronald Numbers and Charles Rosenberg, eds., *The Scientific Enterprise in America: Readings from Isis* (Chicago: University of Chicago Press, 1996). Quote from Morgan B. Sherwood, *Explorations of Alaska, 1865–1900* (New Haven, CT: Yale University Press, 1965), 49. Russell, *Explorations in the Far North*, iii.

43. Historian John Sandlos details the conservationist and colonialist sentiments that emerged around the muskox in the 1890s and early 1900s, especially in Canada. John Sandlos, *Hunters at the Margin: Native People and Wildlife Conservation in the Northwest Territories* (Vancouver: University of British Columbia Press, 2007), 113–38. University of Iowa, Museum of Natural History, "Frank Russell Expedition," available at http://www.uiowa.edu/mnh/researchcollections/russell collection.html, accessed July 13, 2012.

44. Russell, *Explorations in the Far North*, 136–38. See also Frank Russell, "The Journal of Frank Russell," N-2002-037, NWTA, 155–61.

45. Russell, "The Journal of Frank Russell," N-2002-037, NWTA, 158–64.

46. On the expansion of the whaling industry to Baillie Island, see Bockstoce, *Whales, Ice, and Men*, 326–31. See also Ingram and Dobrowolsky, *Waves upon the Shore*, 45–50. On the duties and presence of missionaries and northern police at Herschel Island, see William R. Morrison, *Showing the Flag: The Mounted Police and Canadian Sovereignty in the North, 1894–1925* (Vancouver: University of British Columbia Press, 1985), 77–105, 112–20.

47. For a review of the development of national institutions of science in Canada, see Morris Zaslow, *Reading the Rocks: The Story of the Geological Survey of Canada, 1842–1972* (Toronto: Macmillan Company of Canada, Department of Energy Mines and Resources, 1975). See also Hinckley, *Americanization of Alaska*, 243. Peter Lorenz Neufeld, "De Sainville: Forgotten Mackenzie Mapper," *North* (Winter 1981): 55–57; and Walter Vanast, "Mary Had a Sickly Child: The Social Life of Count de Sainville, a French Aristocrat (Perhaps), at Fort McPherson, 1889–1894," available at http://www.scribd.com/doc/49100414/Sainville-Draft-9b, accessed July 13, 2012.

CHAPTER 2

1. These details come from Canadian Museum of Civilization, "Northern People, Northern Knowledge: The Canadian Arctic Expedition of 1913–1918, by David Gray," http://www.civilization

.ca/cmc/exhibitions/hist/cae/indexe.shtml, accessed December 20, 2011. See also Trevor H. Levere, *Science and the Canadian Arctic: A Century of Exploration, 1818-1918* (New York: Cambridge University Press, 1993), 390-417.

2. On territorial expansion as a way of empire before the Great War, see Neil Smith, *American Empire: Roosevelt's Geographer and the Prelude to Globalization* (Berkeley: University of California Press, 2003). The final cost of the Canadian Arctic Expedition was $519,370.97, as decided by the House of Commons in March of 1920. Found in Kenneth Gordon Chipman Fonds, MG 30 B 66, Vol. 1, File: Scrapbook, 1918-31, LAC.

3. This characterization of expeditionary movements, and my subsequent analysis of these scientific ventures, draws from the work of historian Robert Kohler. See Robert Kohler, "Finders, Keepers: Collecting Sciences and Collecting Practice," *History of Science* 45 (2007): 428-53; and Robert E. Kohler, "History of Field Science: Trends and Prospects," in *Knowing Global Environments: New Historical Perspectives on the Field Sciences*, ed. Jeremy Vetter (New Brunswick, NJ: Rutgers University Press, 2011), 222-23. For a similar treatment with a Canadian focus, see Suzanne Zeller, *Inventing Canada: Early Victorian Science and the Idea of a Transcontinental Nation* (Montreal: McGill-Queen's University Press, 2009).

4. Lisa Bloom, *Gender on Ice: American Ideologies of Polar Expeditions* (Minneapolis: University of Minnesota Press, 1993).

5. Trevor H. Levere, "Vilhjalmur Stefansson, the Continental Shelf, and a New Arctic Continent," *British Journal for the History of Science* 21, no. 2 (1988): 233-47. Clements Markham and Ejnar Mikkelsen, "On the Next Great Arctic Discovery: The Beaufort Sea," *Geographical Journal* 27, no. 1 (1906): 1-11. On phantom islands, see Henry Stommel, *Lost Islands: The Story of Islands That Have Vanished from Nautical Charts* (Vancouver: University of British Columbia Press, 1984).

6. Phillip Buckner, ed., *Canada and the British Empire* (Oxford: Oxford University Press, 2008), 66-107. These two men had met each other on a separate expedition to the Norwegian Arctic in 1901, the Baldwin-Ziegler Expedition, of which Leffingwell was the scientific director. Ernest de Koven Leffingwell, "My Polar Explorations, 1901-1914," *Explorers Journal* 39, no. 3 (1961): 2-14. On the funding for the Anglo-American Expedition, see Ejnar Mikkelsen, *Conquering the Arctic Ice* (Philadelphia: George W. Jacobs, 1909), 1-10.

7. "Report of the Mikkelsen-Leffingwell Expedition," *Bulletin of the American Geographical Society* 39, no. 10 (1907): 607-20.

8. For more on Stefansson's "discovery" of the Blond Eskimos, see Richard Diubaldo, *Stefansson and the Canadian Arctic* (Montreal: McGill-Queen's University Press, 1978), 33-57; and Stefansson, *My Life with the Eskimo* (New York: Macmillan, 1913), 203-304. On the reaction in the print media, see Michael Robinson, "Blonde Eskimos and Yellow Journalism: Reforming the Arctic Narrative in Progressive America," paper presented to the History of Science Society, 1996. My thanks to the author for allowing me to review the text of this presentation. John L. Steckley covers the emergence and settlement of the "Blond Eskimo" in great detail. See John L. Steckley, *White Lies about the Inuit* (Peterborough, ON: Broadview Press, 2008), 77-102.

9. On Darwinism in its historical moment, see Donald Worster, *Nature's Economy: A History of Ecological Ideas* (New York: Cambridge University Press, 1994), 145-69. Steckley, *White Lies about the Inuit*, 101. As historical geographer Bruce Braun has cogently argued, "it is in the continuous *failure* to locate the not yet destroyed" that travelers to colonial outposts came to understand themselves as modern. Landscapes of loss, Braun says, provided evidence of "modernity's destructive force (look it's already happened here!)" and compelled "additional rounds of nostalgic yearnings." Bruce Braun, *The Intemperate Rainforest: Nature, Culture, and Power on Canada's West Coast* (Minneapolis: University of Minnesota Press, 2002), 110-12.

10. On the orders for the Canadian Arctic Expedition, see "Memorandum for Instructions, re: Arctic expedition, by O.E.L (n.d.)," Rudolph Martin Anderson Fonds, MG 30 40, Vol. 10, File 3, LAC.

11. Lewis Green, *The Boundary Hunters: Surveying the 141st Meridian and the Alaska Panhandle* (Vancouver: University of British Columbia Press, 1982), 159-68. Kurkpatrick Dorsey, *The Dawn of Conservation Diplomacy: U.S.-Canadian Wildlife Protection Treaties in the Progressive Era* (Seattle: University of Washington Press, 1998), 8-11. Bruce W. Hodgins and Gwyneth Hoyle, *Canoeing North into the Unknown: A Record of River Travel, 1874-1974* (Toronto: Dundurn, 1997), 132.

12. For one review of the Canadian takeover of the Second Stefansson-Anderson Expedition, see D. Le Bourdais, *Stefansson: Ambassador of the North* (Montreal: Harvest House, 1963), 60-70. Grosvenor, the director of the National Geographic Society, issued a statement to the press to quell any rumors of friction between the two countries, noting that his institution was interested in the scientific results only, not political claims to territory. Cable from Grosvenor to Stefansson, February 15, 1913, Rudolph Martin Anderson Fonds, MG 30 40, Vol. 1, File: Correspondence, Jan-Feb 1913, LAC.

13. Janice Cavell and Jeff Noakes, *Acts of Occupation: Canada and Arctic Sovereignty, 1918-1925* (Seattle: University of Washington Press, 2011), 9.

14. On the Dominion Government Expedition of 1903-4, see W. Gillies Ross, "Canadian Sovereignty in the Arctic: The *Neptune* Expedition of 1903-1904," *Arctic* 29, no. 2 (1976): 87-104. On Canadian claims to the Arctic archipelago in the first decade of the 1900s, see Janice Cavell, "'As Far as 90 North': Joseph Elzear Bernier's 1907 and 1909 Sovereignty Claims," *Polar Record* 46, no. 4 (2010): 372-73. On Chipman's role, see Letter from Commissioner of Customs to KG Chipman, May 28, 1913, Kenneth Gordon Chipman Fonds, 1913-18, MG 30 B 66, Vol. 1, File: Correspondence and Reports, LAC. *Burwick Register*, March 13, 1913 (found in Kenneth Gordon Chipman Fonds, 1913-18, MG 30 B 66, Vol. 2, File 1: Scrapbook No. 1. 1913).

15. The Royal Canadian Navy, which was established in 1910, had six vessels, several of which were former British naval ships. The manuscript notes, photographic material, and journals of the scientists were to be returned to the Geological Survey and the Department of Naval Service, the venture's two leading agencies. See "Memorandum for Instructions re: Arctic expedition, by O. E. L (O. E. LeRoy), n.d.," Rudolph Martin Anderson Fonds, MG 30 40, Vol. 10, File 3, LAC. The control of personal diaries was a matter of great contention in the Canadian Arctic Expedition. See Diubaldo, *Stefansson and the Canadian Arctic*, 76-125.

16. On Greenland, see *New York Times*, February 10, 1906, and "The Crocker Land Expedition," *Bulletin of the American Geographical Society* 44, no. 3 (1912): 189-93. On Alaskan interior, see Mary C. Rabbit, *Minerals, Lands, and Geology for the Common Defence and General Welfare: A History of Public Lands, Federal Science and Mapping Policy, and Development of Mineral Resources in the United States*, vol. 3, *1904-1939* (Washington, DC: US Geological Survey, 1986), 2-65; and Morgan B. Sherwood, *Explorations of Alaska, 1865-1900* (New Haven, CT: Yale University Press, 1965), 132-81. See also A. J. Collier, "Geology and Coal Resources of the Cape Lisburne Region, Alaska," Bulletin no. 278 (Washington, DC: US Geological Survey, 1906). After the peak of activity in the Klondike, prospectors and miners fanned out throughout the Yukon River basin, testing little-known areas for possible mother lodes. The reconnaissance of these miners was largely responsible for subsequent gold rushes in Nome (1900) and Fairbanks (1902).

17. Phillip S. Smith and J. B. Mertie Jr., "Geology and Mineral Resources of Northwestern Alaska," Bulletin no. 815 (Washington, DC: US Department of the Interior, Geological Survey, 1930), 7. See Alfred Brooks and others, "Report on the Progress of Mineral Investigations in Alaska, 1906," Bulletin no. 314 (Washington, DC: Department of the Interior: US Geological Survey, 1907), 11-18.

18. G. M. Dawson, "On Some of the Larger Unexplored Regions of Canada," a paper presented at the Ottawa Field Naturalists' Association (Ottawa: Ottawa Naturalist, 1890), 2. Historian Morris Zaslow describes how Reginald Brock brought the Geological Survey of Canada into the "modern scientific age." Morris Zaslow, *Reading the Rocks: The Story of the Geological Survey of Canada, 1842-1972* (Toronto: Macmillan, 1975), 264-87. US geologists helped train Geological Survey of Canada staff. See Rudolph M. Anderson Fonds, MG 30 40, Vol. 10, File 5, "Karluk Chronicle," LAC.

19. W. J. McGee, "Explorations in the Far North, by Frank Russell," *American Anthropologist* 1, no. 3 (1899): 568. Arles Hrdlicka, "Explorations in the Far North," *American Naturalist* 33, no. 390 (1899): 514-16. For a review of the literature on professionalization, see Mark Barrow, *A Passion for Birds: American Ornithology after Audubon* (Princeton, NJ: Princeton University Press, 1998), and Margaret Rossiter, *Women Scientists in America: Struggles and Strategies to 1940* (Baltimore: Johns Hopkins University Press, 1995).

20. On the funding for the Anglo-American Expedition, see Ejnar Mikkelsen, *Conquering the Arctic Ice* (Philadelphia: George W. Jacobs, 1909), 1-10. On the relations between publishers and explorers in Arctic exploration, see Janice Cavell, "Arctic Exploration in Print Culture, 1890-1930," *Papers of the Bibliographical Society of Canada* 44, no. 2 (2006): 7-43. On Leffingwell's disgust with the naming of the venture, see Ernest de Koven Leffingwell, "Autobiographical notes for *Encyclopedia Arctica*," PEKL; and Ernest de Koven Leffingwell, "A Communication from Leffingwell," *University of Chicago Magazine*, 76-79, found in PEKL. On science versus exploration at this time, see Michael Robinson, *The Coldest Crucible: Arctic Exploration and American Culture* (Chicago: University of Chicago Press, 2006), 133-58. Explorers also deployed characterizations of science—especially scientists employed by federal agencies—to underscore the unique values of exploration. In a May 1919 interview with the *Christian Science Monitor*, Stefansson said, "What one needs is a scientist of the Darwin type . . . whose mind is open to the truth of every sort. The scientist in the civil service . . . is likely to have every other attribute that you would expect of a man who arrives at his office at 9:15 in the morning and leaves at 4:55 in the afternoon." A clipping from this publication can be found in Rudolph Martin Anderson Fonds, MG 30 40, Vol. 10, File 10: CAE: Misc Memoranda, LAC.

21. Ernest de Koven Leffingwell, "My Polar Explorations, 1901-1914" *Explorers Journal* 39, no. 3 (1961): 2-14. On his relief, see Leffingwell, "Communication from Leffingwell," 76-79. On his academic credentials, see Roza Ekimov, "Bio Bibliography of Ernest de Koven Leffingwell," in "Articles," PEKL. Leffingwell had local assistance while in the field, from Ned Arey, his Inuit wife, and their children. Leffingwell discussed his interactions with Arey's family in correspondence and in an unpublished manuscript. See Letter from Leffingwell to Andree, November 6, 1906, PEKL; "'Anglo-American Polar Expedition,' August, 1907, Flaxman Island, Alaska," PEKL.

22. Rudolph Martin Anderson, "Canadian Arctic Expedition—Preliminary History," Rudolph Martin Anderson Fonds, MG 30 40, Vol. 10, File 4, p. 11, LAC. See also Burt McConnel, "The Aeroplane in Arctic Exploration," *Scientific American* (September 1916). "Names of Members of the Canadian Arctic Expedition, their age, height and weight and the name and address of the proper person to notify in case of accident, etc.," Rudolph Martin Anderson Fonds, MG 30 40, Vol. 10, File 5, LAC. Of the twelve men listed on this sheet, eight were under thirty years of age. Four were under twenty-five. The oldest was thirty-five. On the "high order" of knowledge, see Kenneth Gordon Chipman Fonds, MG 30 B 66, Vol. 1, File: Geological Survey instructions, June 1913, LAC.

23. These whalers-turned-traders include Ned Arey, Jim Allen, Christian Klengenberg, Peter Lopez, Charles Brower, Tom Gordon, Fritz Wolki, and C. T. Pedersen. Their last names grace many families in the western Arctic today. Rob Ingram and Helene Dobrowolsky, *Waves upon the Shore: A Historical Profile of Herschel Island*, prepared for Heritage Branch, Department of Tourism, Government of Yukon Territory, September 1989, 148-67; Murielle Ida Nagy, "Yukon North Slope Inuvialuit Oral History," *Occasional Papers in Yukon History No. 1*, Heritage Branch, Government of the Yukon, 1994, 1-7; and David Libbey and William Schneider, "Fur Trapping on Alaska's North Slope," in *Le Castor Fait Tout: Selected Papers of the Fifth North American Fur Trade Conference*, ed. Bruce G. Trigger, Toby Morantz, and Louise Dechene (Lake Saint Louis Historical Society, 1987), 335-58. For individual accounts of this transition provided by whalers-turned-traders, see Charles D. Brower, *King of the Arctic: A Lifetime of Adventure in the Far North* (London: Robert Hale, 1958), 125-31, and Arthur James Allen, *A Whaler and Trader in the Arctic, 1895 to 1944: My Life with the Bowhead* (Anchorage: Alaska Northwest Publishing, 1978), 180-88.

24. Robert McGhee, *Beluga Hunters: An Archaeological Reconstruction of the History and Culture of the Mackenzie Delta Kittegaryumiut* (Saint Johns: Institute of Social and Economic Research, Memorial University of Newfoundland, 1974), 5. Historian John Bockstoce notes that in 1894-95, more than one hundred of the Inuit from Point Hope (Alaska) and Point Barrow were at Herschel Island. Flaxman Island, Barter Island, Demarcation Point, Clarence Lagoon, and Shingle Point all solidified near the turn of the twentieth century in response to the growing importance of commercial coastal whaling in social and economic life.

25. On relations with whalers, see "Anglo-American Polar Expedition," August, 1907, Flaxman Island, Alaska, "Manuscripts," 1-3, PEKL. On Canadian Arctic Expedition members learning of the Great War, see Museum of Civilization, "Northern People, Northern Knowledge." On field-workers staying up to date with publications, see Letter from Vilhjalmur Stefansson to Clark Wissler, November 22, 1916, PSC, and Letter from Vilhjalmur Stefansson to "Charlie," December 9, 1910, PSC. On mail, see Leffingwell, *The Canning River Region: Northern Alaska*, Professional Paper 109, US Geological Survey, 1919, 68-69. Leffingwell noted to a colleague and friend that he had composed fifty letters in preparation for the withdrawal of the whaling boats. Letter from Leffingwell to "Andree," November 6, 1906, PEKL.

26. Vilhjalmur Stefansson to "Percy," December 12, 1910, PSC. For details on the measurements and collections Stefansson took of "Blond Eskimos," see Vilhjalmur Stefansson to Clark Wissler, December 5, 1910, PSC. See also Vilhjalmur Stefansson to Herman C. Bumpus, January 21, 1910, PSC. On trading for rifles and ammunition, see Vilhjalmur Stefansson to Charles Hamel, January 24, 1911, PSC. On the promise to take collections to Nome in summer 1911, see Rudolph Anderson to Herman C. Bumpus, January 17, 1911, PSC. Stefansson's suspicious "purchase" of ethnological specimens from the Coronation Gulf region is described in Vilhjalmur Stefansson to Herman C. Bumpus, April 19, 1911, PSC. For Stefansson's published words on trading with Natives, see Vilhjalmur Stefansson, "Suitability of Eskimo Methods of Winter Travel in Scientific Exploration," *Bulletin of the American Geographical Society of New York* 60, no. 1 (1908): 3. See also Vilhjalmur Stefansson to Clark Wissler, December 5, 1910, PSC. and Herman C. Bumpus to Vilhjalmur Stefansson, May 12, 1910, PSC. On sending supplies back to funders, see Stefansson, "Suitability of Eskimo Methods of Winter Travel in Scientific Exploration, 210.

27. For a description of Stefansson's return from the north in 1912 and the beginnings of the Canadian Arctic Expedition, see Diubaldo, *Stefansson and the Canadian Arctic*, 66-68. On the international lecture tour, see *London Times*, April 22, 1913. For NGS and AMNH's continued support of Stefansson, see *New York Times*, May 26, 1913.

28. During these travels, Jenness relied on a phonograph to record Native songs and stories. He was unfamiliar with the language of the "Blond Eskimo," even though he had previously studied and become familiar with languages of Inuit along the Beaufort Sea coast. For a description of the scientific staff of the Canadian Arctic Expedition, see *New York Times*, May 21, 1913. On Jenness's duties, see *New York Times*, August 9, 1916.

29. Steckley, *White Lies about the Inuit*, 90-97.

30. Expeditions in the western Arctic were thus comprised of what Arctic historians have called touring scientific "omnivores." Sverker Sörlin, "Rituals and Resources of Natural History: The North and the Arctic in Swedish Scientific Nationalism," in *Narrating the Arctic: A Cultural History of Nordic Scientific Practices*, ed. Michael Bravo and Sverker Sörlin (Canton, MA: Science History Publications, 2002), 109. On the role of go-betweens in the making of colonial science, see Kapil Raj, *Relocating Modern Science: Circulation and the Construction of Knowledge in South Asia and Europe, 1650-1900* (New York: Palgrave Macmillan, 2007).

31. Museum of Civilization, "Northern People, Northern Knowledge."

32. A letter from Anderson describing these reactions, dated October 14, 1908, is found in "News from the Museum's Arctic Explorers," *American Museum Journal* 9, no. 5 (1909): 113. See also Vilhjalmur Stefansson, "My Quest in the Arctic," *Harper's Monthly Magazine*, December 1912, 10; Vilhjalmur Stefansson to Herman C. Bumpus, August 19, 1909, PSC. See also David Gray,

Arctic Shadows: The Arctic Journeys of Dr. R. M. Anderson, DVD, directed by David Gray, Mountain Studios and Grayhound Information Services, 2009. Peter Geller has analyzed the role of pictures of the Canadian north in the federal government's understanding of sovereignty over the region in the early 1900s. Peter Geller, *Northern Exposures: Photographing and Filming the Canadian North, 1920-1945* (Vancouver: University of British Columbia Press, 2006), 18–50.

33. Stuart Jenness, ed., *Arctic Odyssey: The Diary of Diamond Jenness, 1913-1916* (Hull, QC: Canadian Museum of Civilization, 1991), 166-86, 240-60, and 370-90. Reference to "Southern Party Boys" in Rudolph Anderson's diary, February 27, 1916, as cited in Museum of Civilization, "Northern People, Northern Knowledge." See also Letter from Stefansson to Anderson, October 13, 1912. Rudolph Martin Anderson Fonds, MG 30 40, Vol. 1, File 7: Correspondence, LAC.

34. Robert Kohler, *All Creatures: Naturalists, Collectors, and Biodiversity, 1850-1950* (Princeton, NJ: Princeton University Press, 2006), 161.

35. A copy of the contract Jenness signed is reprinted in Stuart Jenness, *Arctic Odyssey*, 728. Appendix 3 in *Arctic Odyssey* offers a complete reproduction of the trade items Diamond Jenness exchanged for his anthropological collections. See Stuart Jenness, *Arctic Odyssey*, 672-96.

36. On interactions with Brower, see Stuart Jenness, *Arctic Odyssey*, 145-47. On purchase of the collection at Point Hope, see ibid., 622. On timing of digs and monthly salary, see ibid., 212-27, 240. In one case, the anthropologist paid assistants with a 20-gauge shotgun, a box of brass shells, $9 cash, a tin of lard, a pound of tobacco, and a tin of baking powder—a bounty whose cash equivalent surpassed Jenness's own monthly salary. Diamond Jenness, "The Eskimos of Northern Alaska: A Study in the Effect of Civilization," *Geographical Review* 5 (1918): 89-101.

37. Richard Finnie made note of Ottawa's change in orientation to northern geography and administration after the Great War in Richard Finnie, *Canada Moves North* (New York: Hurst and Blackett, 1942), 30. On the end of territorial expansion as a model of empire after the Great War, see Smith, *American Empire*, 1-30. Peder Roberts and Lize-Marie van der Watt suggest that, after the Canadian Arctic Expedition, this type of "grand" expedition petered out, giving way to smaller, more "prosaic" ventures. See Peder Roberts and Lize-Marie van der Watt, "On Past, Present, and Future Arctic Expeditions," in *The New Arctic*, ed. Birgitta Evengård, Joan Nymand Larsen, and Øyvind Paasche (Cham, Switzerland: Springer, 2015), 61.

38. Details of the collections come from Canadian Museum of Civilization, "Northern People, Northern Knowledge: The Canadian Arctic Expedition of 1913-1918, by David Gray." Stuart E. Jenness, *Stefansson, Dr. Anderson, and the Canadian Arctic Expedition, 1913-1918* (Gatineau, QC: Canadian Museum of Civilization, 2011), 307-10.

39. The distribution of Canadian Arctic Expedition reports is mentioned in a note written by Rudolph Anderson on July 4, 1927. Rudolph Martin Anderson Fonds, MG 30 40, File 13: "Memoranda and Reports about the Canadian Arctic Expedition," LAC. A. W. Greeley, *The Polar Regions in the Twentieth Century: Their Discovery and Industrial Evolution* (Boston: Little, Brown, 1928), 73. Brooks's endorsement is found in Leffingwell, *Canning River Region*, 9. On mineral withdrawals in Canada, see Letter from J. J. O'Neill to R. G. McConnell, March 3, 1919, in J. J. O'Neill Fonds, MG 30 B 171, Vol. 1, File 14, LAC. O'Neill wrote that Order in Council of December 21, 1918 (PC 5154) withdrew "that part of arctic Canada lying between 105 and 116 degrees west longitude and north of 165 degrees north latitude." For mineral withdrawals in Alaska, and Leffingwell's role in them, see Phillip S. Smith and J. B. Mertie Jr., *Geology and Mineral Resources of Northwestern Alaska*, Bulletin no. 815 (Washington, DC: US Geological Survey, 1930), 274-75. Geller discusses the importance of pictures in the twinned projects of Arctic science and Arctic sovereignty, especially sharing and circulating them. See Geller, *Northern Exposures*, 19-21.

40. On Anderson's involvement in conservation, see John Sandlos, *Hunters at the Margin: Native People and Wildlife Conservation in the Northwest Territories* (Vancouver: University of British Columbia Press, 2007). Anderson's lone biographer has called him "the greatest mammalogist that Canada has ever had." The quotation is from Gray, *Arctic Shadows*.

41. For this work, he was referred to as the "Father of Eskimo Archaeology." On Jenness's

subsequent influence in Canada, see Peter Kulchyski, "Anthropology in the Service of the State: Diamond Jenness and Canadian Indian policy," *Journal of Canadian Studies* 28, no. 2 (1993): 21-50; Henry Collins, "Diamond Jenness: Arctic Archaeology," *Beaver* (Autumn 1967): 78-79; Nansi Swazye, *Canadian Portraits: Jenness, Barbeau, Wintemberg; The Man Hunters* (Toronto: Clarke, Irwin, 1960), 40-94.

42. On Stefansson, see Cavell and Noakes, *Acts of Occupation*, 1-11. On Leffingwell's move to California and reflections about the uses of his science, see Ernest de Koven Leffingwell, "My Polar Explorations, 1901-1914," *Explorers Journal* 39, no. 3 (1961): 8-14. Notably, neither Anderson nor Stefansson produced an official report for the Canadian Arctic Expedition.

43. Charles Arnold, Wendy Stephenson, Bob Simpson, and Zoe Ho, eds., *Taimani: At That Time; Inuvialuit Timeline Visual Guide* (Inuvik, NT: Inuvialuit Regional Corporation, 2011), 88.

44. Ishmael Alunik, Eddie D. Kolausok, and David Morrison, *Across Time and Tundra: The Inuvialuit of the Western Arctic* (Seattle: University of Washington Press, 2003), 122-23. On Patsy Klegenberg's work with HBC, see "Western Arctic District," *Beaver* (June 1934): 60. There are also great descriptions of these Inuit-led changes in the Museum of Civilization exhibit of the Canadian Arctic Expedition. Canadian Museum of Civilization, "Northern People, Northern Knowledge: The Canadian Arctic Expedition of 1913-1918, by David Gray." For an insightful analysis of Stefansson's anthropology and scientific representations in light of his personal relationships in the western Arctic, see Gísli Pálsson, ed., *Travelling Passions: The Hidden Life of Vilhjalmur Stefansson* (Hanover, NH: Dartmouth College Press, University Press of New England, 2005).

45. Arnold et al., *Taimani*, 88. See also Albert Elias and Charles Arnold, "The Schooner Era in Twentieth-Century Inuvialuit History," a presentation before the 18th Inuit Studies Conference, October 26, 2012. Don Bisset, "Lower Mackenzie Region: An Economic Survey," Industrial Division: Department of Indian Affairs and Northern Development, October, 1967, 47-48. American C. T. Pedersen, a former whaler who started the CanAlaska Trading Company in the early 1920s, supplied Mackenzie Delta Inuit with schooners crafted in the south. By doing so, he could avoid paying Canadian customs by traveling no farther than Herschel Island. But he also realized Inuit skill and interest in trapping and preparing fox. On Pedersen's tactics, see "Interview of Mrs. Marjory Robertson," N-1998-042-001, NWTA.

46. A. E. Porsild, "Field Journal of an Expedition through Alaska Yukon and the Mackenzie District being a botanical reconnaissance with special reference to the suitability of the country for domesticated reindeer. Also many notes on the physiography of the country, its inhabitants, wild life and general economic conditions, 1926-1928," Alf Erling Porsild Fonds, R5495-0-7-E, M-1958, LAC.

CHAPTER 3

1. Epigraph from A. E. Porsild, "Trip to Alaska to Select Reindeer to Be Purchase for Delivery to the Mackenzie Delta, NWT, Autumn and Winter, 1929-1930. Daily Journal of A. E. Porsild," 39, Alf Erling Porsild Fonds, Microfilm reel M-1958, LAC.

2. Porsild's involvement with the Canadian Reindeer Project is given in Alf Erling Porsild, *Reindeer Grazing in Northwest Canada: Report of an Investigation of Pastoral Possibilities in the Area from the Alaska-Yukon Boundary to Coppermine River* (Ottawa: F. A. Acland, 1929), 6-14, 29. A useful and comprehensive treatment of Porsild's research and correspondence during his "reindeer years" is offered by Wendy Dathan, *The Reindeer Botanist: Alf Erling Porsild, 1901-1977* (Calgary: University of Calgary Press, November 2012).

3. R. M. Hill, "Mackenzie Reindeer Operations," Northern Coordination and Research Centre, Department of Indian Affairs and Northern Development, August 1967, DHC. On population estimates, J. Sonnenfeld, "An Arctic Reindeer Industry: Growth and Decline," *Geographical Review* 49,

no. 1 (1959): 77-78. See also A. D. Johnson, "Brief History of the Reindeer in the Arctic," found in RG 75, Box "Historical Files, 1929-1948," Folder "History—Reindeer in Barrow Region, 1942," NARA-A. The Canadian Reindeer Project did not end with World War II or even 1959, as is implied here. It continued to be passed back and forth between private and public hands throughout the 1960s and 1970s. As of fall 2015, the small extant herd in the Mackenzie Delta is owned partly by a private individual and partly by the Inuvialuit Regional Corporation.

4. The initial Canadian herd remained small and isolated compared to what became of the Canadian Reindeer Project in the 1930s and 1940s. See C. L. Andrews, *The Eskimo and His Reindeer in Alaska* (Caldwell, ID: Caxton Printers, 1939), 30-37; Gilles Seguin, "Reindeer for the Inuit: The Canadian Reindeer Project, 1929-1960," *Muskox* 38 (1991): 1-10; Roxanne Willis, "A New Game in the North: Native Reindeer Herding, 1890-1940," *Western Historical Quarterly* 37 (Autumn 2006): 277-301; and John Sandlos, "Where the Reindeer and Inuit Should Play: Animal Husbandry and Ecological Imperialism in Canada's North," unpublished manuscript. My thanks to the author for allowing me to review this piece.

5. Carl J. Lomen, *Fifty Years in Alaska* (New York: David McKay, 1954), 117-35, 144-53, and 170-81. See also Vilhjalmur Stefansson, "Arctic Headlines in Fact and Fable," *Foreign Policy Association* 51 (January 1945): 79. Nelson also hired veterinary scientist Seymour Hadwen to assist early operations in Alaska.

6. This was the subtitle to the 1922 report of the Department of the Interior, Reindeer and Muskox: Report of the Royal Commission upon the Possibilities of the Reindeer and Musk-Ox Industries in the Arctic and Sub-Arctic Regions (Ottawa, 1922). The list of witnesses can be found in Department of the Interior, *Reindeer and Muskox*, 9-11.

7. Historian John Sandlos has argued the discursive practices relating to conservation in the north created an "Arctic Pastoral," in which bureaucrats, sportsmen, scientists, and other conservationists positioned the Native hunter as "irrational and destructive" and portrayed the Arctic tundra as an environment ripe for government-sponsored development. Certainly, Sandlos's argument applies to US bureaucrats and businessmen as well. For an overview of the "Arctic Pastoral" concept, see John Sandlos, *Hunters at the Margin: Native People and Wildlife Conservation in the Northwest Territories* (Vancouver: University of British Columbia Press, 2007), 161-70.

8. Randal Pokiak, *Inuvialuit History*, unpublished manuscript, 1991, 58, ICRC. The formal designation of Inuit as wards of the state did not occur until an amendment to the Indian Act made in 1924. This amendment appears to have been an attempt to legitimize federal appropriations for forms of relief that were distributed to Inuit through missions and fur posts. See John Leonard Taylor, *Canadian Indian Policy during the Inter-War Years, 1918-1939* (Ottawa: Department of Indian Affairs and Northern Development, 1984), 87-88. On troubles with enforcement, see Sandlos, *Hunters at the Margin*, 126.

9. Pokiak, *Inuvialuit History*, 44-58, quote from page 58. There are many other accounts of Mangilaluk in oral histories. See Elisa J. Hart, *Reindeer Days Remembered* (Inuvik, NT: Inuvialuit Cultural Resource Center, 2001), 14.

10. See "Royal Commission: Reindeer and Muskox Industry, Vol. 1," RG 33-105, LAC.

11. Notes from Jenness's lecture are found in "A Lecture Delivered at the Arts and Letters Club, by Diamond Jenness, Victoria Memorial Museum, Jan 9, 1923: 'Our Eskimo Problem,'" Rudolph Martin Anderson Fonds, MG 30 40, Vol. 14, File 1: Eskimos, LAC. Janice Cavell and Jeff Noakes, *Acts of Occupation: Canada and Arctic Sovereignty, 1918-25* (Vancouver: University of British Columbia Press, 2010), 9. On post-World War I development politics in an imperial context, see Neil Smith, *American Empire: Roosevelt's Geographer and the Prelude to Globalization* (Berkeley: University of California Press, 2003), xviii-15.

12. The records of the hearings are found in Library and Archives Canada, RG 33 105, under the titles "Royal Commission: Reindeer and Muskox Industry, Vol. 1," and "Royal Commission: Reindeer and Muskox Industry, Vol. 2." The quotation is from Bishop Isaac Stringer from the

February 4, 1920, meeting, found on pages 197-98 of the Vol. 1 hearings. Department of the Interior, *Reindeer and Muskox*, 12-24, 45, 95.

13. Seguin, "Reindeer for the Inuit," 1-10.

14. Stefansson's controversial nature has been analyzed by his many biographers. See Richard Diubaldo, *Stefansson and the Canadian Arctic* (Montreal: McGill-Queen's University Press, 1978); and William Hunt, *Stef: A Biography of Vilhjalmur Stefansson, Canadian Arctic Explorer* (Vancouver: University of British Columbia Press, 1986). Department of the Interior, *Reindeer and Muskox*, 26.

15. O. S. Finnie to Gibson, January 1926, as quoted in Patricia Wendy Dathan, "The Reindeer Years: Contribution of A. Erling Porsild to the Continental Northwest" (master's thesis, Department of Geography, McGill University, 1988). For quotation at end of paragraph, see Letter from O. S. Finnie to Rev. Canon C. W. Vernon, June 10, 1929, Microfilm Reel T-13267, Vol. 759, File 4824—WT Lopp, 1925-39, LAC. See also Letter from O. S. Finnie to W. H. Collins, May 14, 1928, RG 132, Vol. 23, File 364, LAC; Letter from O. S. Finnie to M. O. Malter, February 20, 1928, RG 132, Vol. 23, File 364, LAC.

16. The details of this meeting among US and Canadian officials are drawn from several sources. See Letter to A. E. Porsild, March 23, 1926, RG 132, Vol. 31, File 4492, LAC. See also Letter from O. S. Finnie to Colonel Starnes, May 19, 1926, Microfilm Reel #T13273, Vol. 765, File 5095—Porsild, 1926-36, LAC. On requirements of applied botanist, see O. S. Finnie to Moran, December 4, 1925, Microfilm Reel #T-13267, Vol. 759, File 4824—WT Lopp, 1925-39, LAC. On hiring a "Botanist" and "Assistant Botanist," see O. S. Finnie to Moran, April 20, 1926, #T13273, Vol. 765, File 5095—Porsild, 1926-36, Vol. 765, File 5095—Porsild, 1926-36, LAC. Industrialization of reindeer herding in other contexts typically followed changes in transportation technology. See Peter Sköld, "Perpetual Adaptation? Challenges for the Sami and Reindeer Husbandry in Sweden," in *The New Arctic*, ed. Birgitta Evengård, Joan Nymand Larsen, and Øyvind Paasche (Cham, Switzerland: Springer, 2015), 47-49.

17. Porsild, *Reindeer Grazing in Northwest Canada*, 5-6. Letter from Vilhjalmur Stefansson to Carl J. Lomen, June 20, 1927, Lomen Family Papers, Series 2, Box 10, Folder 232, APR-UAF. See also Percy Cox, Professor Seward, and Colonel Vanier, "The Reindeer Industry and the Canadian Eskimo: Discussion," *Geographical Journal* 88, no. 1 (1936): 17-19. Dathan provides a rich description of the Porsild's upbringing. See Dathan, "Reindeer Years," 2-10. This situation echoes the complicated emergence of grazing sciences in America and Canada over the turn of the twentieth century. See John Sandlos, "Where the Scientists Roam: Ecology, Management, and Bison in Northern Canada," in *Canadian Environmental History: Essential Readings*, ed. David Freeland Duke (Toronto: Canadian Scholars' Press, 2006), 333-60; Morgan Sherwood, *Big Game in Alaska* (New Haven, CT: Yale University Press, 1981); Christian C. Young, "Defining the Range: The Development of Carrying Capacity in Management Practice," *Journal of the History of Biology* 31 (1998): 61-83.

18. Quoted in Dathan, "Reindeer Years," 49.

19. Ibid.

20. "Record of Members: Society of American Foresters," L. J. Palmer Papers, Box 2, Folder 1, APR-UAF.

21. L. J. Palmer and Seymour Hadwen, *Reindeer in Alaska*, Bulletin no. 1089, September 22, 1922 (Washington, DC: US Department of Agriculture, 1926), 19-23. See also L. J. Palmer, *Progress of Reindeer Grazing Investigations in Alaska*, Bulletin no. 1423, October 1926 (Washington, DC: US Department of Agriculture, 1927), 1-11; and L. J. Palmer to E. W. Nelson, March 22, 1923, Lawrence J. Palmer Papers, APR-UAF.

22. Palmer, *Progress of Reindeer Grazing Investigations in Alaska*, 30-31. Palmer determined the nutritive quality of lichens by sending samples to the Bureau of Chemistry and to the Bureau of Plant Industry in Washington, DC. See Letter from L. J. Palmer to Chief of Bureau, Biological Survey, November 6, 1929, RG 75, Bureau of Indian Affairs, Lawrence J. Palmer Correspondence, Box 2, Digestion Studies, 1920-34, NARA-A. "'Progress Report: Quadrat Studies,' Nov 1, 1923," RG 75, Box 4, NARA-A.

23. The Forest History Society, "Range Management Bulletin," http://www.foresthistory.org /ASPNET/Policy/Grazing/Jardinebook.aspx, accessed March 8, 2013; and The Forest History Society, "Letter," http://www.foresthistory.org/ASPNET/Policy/Grazing/GrazingInspection.aspx, accessed March 8, 2013. See also the Forest History Society, "Range Management and Research: 1910 to 1929," http://www.foresthistory.org/ASPNET/Publications/region/4/history/chap5.aspx, accessed March 8, 2013.

24. L. J. Palmer, "Progress of Reindeer Grazing Investigations in Alaska," 3, 33. See also Letter from E. W. Nelson to Edmund Seymour (President of American Bison Society), December 30, 1921, RG 22, Entry 162, Box 21, Folder: Experimental Station, Lomen Brothers, General Correspondence, NARA.

25. Willis, "New Game in the North," 293-96. See also US Department of the Interior, Hearings of the Reindeer Committee in Washington, DC, February–March, 1931 (Washington, DC: Office of the Secretary, 1931), 27.

26. W. T. Lopp, "Comparison of Herding Systems Used in the Alaska Reindeer Service," May 16, 1936, in Kathleen Lopp Smith Family Papers, Box 2, Folder 22, UAF-APR.

27. Letter from E. W. Nelson to Edmund Seymour (President of American Bison Society), December 30, 1921, RG 22, Entry 162, Box 21, Folder: Experimental Station, Lomen Brothers, General Correspondence, NARA. See W. B. Bell to Mr. Wm B. Miller, November 27, 1928, Lawrence J. Palmer Papers, APR-UAF. See also "Survey of Conditions of the Indians in the United States: Hearings before a Subcommittee of the Committee on Indian Affairs," US Senate, 74th congress, second session, part 36, Alaska (including reindeer), US Government Printing Office, 1939, 20173. Here, former Alaska Native Service staff member C. L. Andrews testified that the Bureau of Biological Survey was in league with the Lomen Corporation and that scientific research had been marshaled in the service of the state and capital.

28. James C. Scott, *Seeing Like a State: How Certain Schemes to Improve the Human Condition Have Failed* (New Haven, CT: Yale University Press, 1999), 1-9.

29. Quote about files in A. E. Porsild, "Field Journal of an Expedition through Alaska Yukon and the Mackenzie District being a botanical reconnaissance with special reference to the suitability of the country for domesticated reindeer. Also many notes on the physiography of the country, its inhabitants, wild life and general economic conditions. 1926-1928, by A. E. Porsild," Microfilm Reel M-1958, Library and Archives Canada, 8. See also Dathan, "Reindeer Years," 53-62. See L. J. Palmer to Chief of Bureau Biological Survey, August 27, 1926, Lawrence J. Palmer Papers, Box 3, Folder 4, APR-UAF. A. E. Porsild, "Field Journal of an Expedition through Alaska Yukon and the Mackenzie District," 57, Microfilm Reel M-1958, LAC. Porsild, *Reindeer Grazing in Northwest Canada*, 14.

30. Erling Porsild, "The Reindeer Industry and the Canadian Eskimo." *Geographical Journal* 88, no. 1 1936): 6-10. See also Dathan, "Reindeer Years," 118.

31. Porsild, "Reindeer Industry and the Canadian Eskimo," 16.

32. Porsild, *Reindeer Grazing in Northwest Canada*, 29-32. See also Letter from A. E. Porsild, June 25, 1927, Microfilm Reel M-1958, LAC. The letter was not listed to anyone. A. E. Porsild, "Field Journal of an Expedition through Alaska Yukon and the Mackenzie District," 70, Microfilm Reel M-1958, LAC.

33. Dick North, *Arctic Exodus: The Last Great Trail Drive* (Guilford, CT: Lyons Press, 2005), 170-72; and Richard Finnie, *Canada Moves North* (New York: Hurst and Blackett, 1942), 65-70. Finnie and Cory resigned at the end of 1931; the Northwest Territories and Yukon Branch was abolished and territories came under the jurisdiction of the Dominion Lands Board. In subsequent memoranda, administrators expressed the view that, since such a large amount of public funds had been invested in the reindeer experiment over the 1920s, continued expenditures in the post-Depression era were justified.

34. Sami families were promised payment of $60 a month, as well as medical aid, education for children, food rations, and a yearly bonus for growth in the size of the herd. The Sami herders were

also guaranteed private living quarters. In contrast, the foreman was paid $85 per month and had similar benefits, while Inuit apprentices earned $25 per month and had only some of the benefits. On the existing system of contracting between farmers and reindeer herdsmen in Lapland, see Sköld, "Perpetual Adaptation?," 46.

35. Minutes from a meeting of the Interdepartmental Reindeer committee, January 18, 1933, G79–069, File 1–3, NWTA. See also Porsild, "To the Mackenzie Delta to inspect and report on Reindeer Experiment, June–August 1947, Daily Journal of A. E. Porsild," Microfilm M-1958, LAC, 3.

36. Letter from A. E. Porsild, June 16, 1927, Microfilm Reel M-1958, LAC. Porsild wrote, "The case of Tarpoq is not unique. Similar cases have been met with in other places in Alaska, when an eskimo, though a good reindeer man under the supervision of a white man, when left to himself, soon starts to neglect his herd when his increase and profits is not up to his expectations." See Memo from R. A. Gibson to J. A. Parsons, September 28, 1942, RG 109, Vol. 485, File 1, LAC. This same file includes a number of "Information Reports" that profile the Inuit herders recruited to the Reindeer Project and Stanley Mason.

37. "Minutes of a Meeting of the Inter-Departmental Reindeer Committee held on Friday, 11th October, 1935, at 2:00 P.M., in Room 604, Norlite Building," RG 85, File 11, Pt. 1, LAC. "Minutes of a Meeting of the Inter-Departmental Reindeer Committee held Tuesday, 14th January, 1941, at 3:00 P.M. in Room 801, Norlite Building," RG 85, File 11, Pt. 2, LAC. For a summary of controversies among scientists, Inuit, and Sami along the drive, see North, *Arctic Exodus*, 195–97, 217.

38. List for File: Staff at Reindeer Station, Nov. 15, 1935, Microfilm reel, T-13323, Vol. 822, File 7128, Lapp herders, 1929–1938, LAC. Lands, Parks, and Forests Branch, *Canada's Reindeer*, Northwest Territories Administration (Ottawa, 1940), 5–7. Finnie, *Canada Moves North*, 141. See also A. E. Porsild, 1947 Report, 1–7, RG 109, Vol. 491, File 111, LAC. Porsild notes that between 1936 and 1947, the Canadian Reindeer Project sent Aklavik 5,000 carcasses, amounting to 500,000 pounds of meat.

39. Willis, "New Game in the North," 297–300. See also Dean F. Olson, *Alaska Reindeer Herdsmen: A Study of Native Management in Transition* (Fairbanks, AK: Institute of Social, Economic, and Government Research, 1969). See especially chapter 3: "The Period of Decline, 1933-1950," http://www.alaskool.org/projects/reindeer/history/iser1969/reindeer_3.html, accessed March 8, 2013.

40. Sonnenfeld, "Arctic Reindeer Industry," 82–85. Chris Southcott identifies the interwar period as a moment when US and Canadian officials confronted the tensions of national development and demands of global capital in the Arctic. See Chris Southcott, "History of Globalization in the Circumpolar World," in *Globalization and the Circumpolar North*, ed. Lassi Heininen and Chris Southcott (Fairbanks: University of Alaska Press, 2010), 44.

41. Hart, *Reindeer Days Remembered*, 25. David Roland said, "I could have been a herder when I was a young man, but when the fur was good, I didn't like to have a boss. Yes, then the fur [prices] got so low that I had to take a job. That was in 1953." Medical Health Officer to R. A. Gibson, September 8, 1938, RG 85, File 28, Pt. 2, LAC.

42. Minutes of the Interdepartmental Reindeer Committee from June 18, 1935, Microfilm reel T-13324, Vol. 82, File 7128, "Lapp herders, 1929–1938," LAC. In March 1935, alone, more than twenty-five trappers from the Mackenzie Delta, the majority of whom were Inuit, applied for permits to enter the Reindeer Grazing Reserve in order to trap. See A. E. Porsild to J. Lorne Turner, Esq., April 3, 1935, RG 85, File 28: Permits for Reindeer Reserve, 1935-37, LAC. The policy of allowing permits to all Natives is further detailed in Medical Health Officer to R. A. Gibson, September 8, 1938, RG 85, File 28, Pt. 2, LAC.

43. Rob Ingram and Helene Dobrowolsky, *Waves upon the Shore: A Historical Profile of Herschel Island*, prepared for Heritage Branch, Department of Tourism, Government of Yukon Territory, September 1989, 164–67. Plans to build up Tuktoyaktuk were in place as early as 1930. F. H. Kitto,

"The Northwest Territories, 1930" (Department of the Interior, Northwest Territories and Yukon Branch, 1933), 38-45.

44. Joseph Sonnenfeld, "Changes in Subsistence among the Barrow Eskimo" (PhD diss., Johns Hopkins University, 1956), 307-9. A. D. Johnson, "Brief History of the Reindeer in the Arctic," found in RG 75, Box "Historical Files, 1929-1948," Folder "History—Reindeer in Barrow Region, 1942," NARA-A.

45. Letter from Pete Paulson to Mrs. Isabella Hutchinsen, n.d., Dorothy and Grenold Collins Collection, Box 7, Series 5, Folder 1, UAA. Paulson writes that after Barter Island and Colville residents slaughtered herds, most families living along the northeastern slope moved either to Barrow (and on to Wainwright) or into Canada. Royal Canadian Mounted Police memo, "Alaskan Natives entering Canada at Demarcation Point, Y.T., Trapping in N.W.T.," July 23, 1946, found in Microfilm reel T-13329, Vol. 829, File 7282: "Mackenzie Delta Game Preserve, 1930-1949 (Parts 1 and 2)," LAC. The Hudson's Bay Company stood to gain from this situation, having access to more trappers given both the removal of Pedersen's competition and the addition of a skilled workforce. The Department of Mines and Resources did respond in 1938 by discontinuing trapping by white men in the Northwest Territories if they did not already have a permit or license and did not reside in the area. See R. A. Gibson to W. A. Parsons, March 10, 1939, RG 85, File 28, Pt. 2: Permits for Reindeer Reserve, 1938-45, LAC.

46. R. H. G. Bonnycastle, "Canada's Reindeer Experiment," Proceedings of the North American Wildlife Conference, February 3-7, 1936, Senate Committee Print, 74th congress, 2nd session (Washington, DC, 1936), 424-27. Porsild, Reindeer Industry and the Canadian Eskimo, 4. On "easily trained," see General Foreman to R. A. Gibson, April 1, 1942, in RG 85, File 32: Annual Reports, 1939-47. See also J. T. Parsons, General Foreman, Government Reindeer Depot to Mr. R. A. Gibson, Director, Lands, Parks and Forest Branch, September 1, 1940, RG 109, Vol. 485, File 1, LAC. On the composition of Native herders, see Hart, Reindeer Days Remembered, 19-28, 72-79. On summer school, see "Annual Report: Reindeer Station, N.W.T.: April 1, 1952-March 31, 1953," RG 85, File 88, LAC.

47. Hart, Reindeer Days Remembered, 19-28, 72-79. "Disaster at Tuktuk," Moccasin Telegraph (March 1945); 24. Available in the Public Reference file on Reindeer, NWTA. A sixth Native herd, the fourth to be started after the Cally accident, began in 1954 and was returned in 1964. It was the longest-running Native herd at that point.

48. Pokiak, Inuvialuit History, 80. Elisa Hart and Cathy Cockney, Yellow Beetle Oral History and Archaeology Project (Inuvialuit Social Development Program, 1999), 1. For a compelling history of Inuvik's construction in the 1950s and its effect on local labor and land use, see Dick Hill, Inuvik: A History, 1958-2008; The Planning, Construction, and Growth of an Arctic Community (Victoria, BC: Trafford Publishing, 2008), 38-59. See also Hart, Reindeer Days Remembered, 84. On labor pool problems in conjunction with DEW line and Inuvik's construction, see "Annual Report, Reindeer Station, April 1, 1958-March 31, 1959," G1979-003-70-1, NWTA. On Lloyd Binder, see Andrew Livingstone, "Last of the Original Reindeer Herders," Northern News Services online, http://www .nnsl.com/frames/newspapers/2015-03/mar5_15rei1.html, accessed December 3, 2015. In April 2011, I had the chance to visit the reindeer herd and Reindeer Station to learn more about the settlement and herding practices. My thanks to Dave Halpine, Henrik, and Meltzer Sydney.

49. "Extracts from Report of W. B. Miller on Kuskokwim Reindeer Investigations, Dated Jan 17, 1931," an addendum within Reindeer Hearings, Washington, DC, 1931.

50. On published analyses of the Canadian Reindeer Project, see Sonnenfeld, "Arctic Reindeer Industry," 77-94; George W. Scotter, "Reindeer Husbandry as a Land Use in Northwestern Canada," in Proceedings of the Productivity and Conservation in Northern Circumpolar Lands Conference, ed. W. A. Fuller and P. G. Kevan. IUCN, n.s., 16 (1969): 159-69. Unpublished memos and annual reports from Reindeer Station post managers between 1952 and 1959 detail many of these "problems" with reindeer and make gestures to research on the Alaskan industry's failure. See, for

example, L. A. C. O. Hunt, "Some Observations on the Reindeer Industry, 1952," G1979-003-70-1, NWTA; and Letter from R. C. Robertson to "Mr. Steele," May 27, 1960, G1979-003-70-1, NWTA. These documents gave little consideration to the ways Inuit residents consciously avoided the program.

CHAPTER 4

1. Stephen J. Eszenyi, Officer in Charge of US Naval Arctic Test Station to Office in Charge, US Naval Advance Base Depot, "Monthly Report no. 1-47," 6-22, July 16, 1947, found in RG 57, Entry: Geologic Division, Records of the Alaska Terrain and Permafrost Section, 1945-54, Box 2, Folder: Arctic Research Laboratory, Data on Barrow, Alaska, NARA.

2. Ibid., 24.

3. Richard Finnie, *CANOL* (San Francisco: Taylor and Taylor, 1945).

4. The Canadian Encyclopedia, "Northwest Staging Route," written by Kenneth S. Coates, http:// www.thecanadianencyclopedia.com/articles/northwest-staging-route, accessed March 12, 2013. Wikipedia contributors, "Canol Road," Wikipedia, the Free Encyclopedia, http://en.wikipedia.org /wiki/Canol_Road, accessed March 12, 2013. See also Kenneth S. Coates and William R. Morrison, "Soldier-Workers: The U.S. Army Corps of Engineers and the Northwest Defense Projects, 1942-1946," *Pacific Historical Review* 62, no. 3 (1993): 281-83.

5. The Canadian Encyclopedia, "Canol Pipeline," written by Kenneth S. Coates, http://www .thecanadianencyclopedia.com/articles/canol-pipeline, accessed March 12, 2013.

6. US Department of Energy, "Naval Petroleum Reserves-Profile," updated on December 22, 2011, http://fossil.energy.gov/programs/reserves/npr/, accessed March 12, 2013. Quote about Public Land Order 82 from Peter A. Coates, *The Trans-Alaska Pipeline Controversy: Technology, Conservation, and the Frontier* (Anchorage: University of Alaska Press, 1993), 99.

7. *New York Times*, November 5, 1950. H. E. Landsberg, "Alaska Research and National Defense," in Naval Arctic Research Laboratory Records, Box 74, Folder: "Alaska Science Conference, First, November 1950," APR-UAF. Landsberg was the executive director of the Committee on Geophysics and Geography for the US Research and Development Board.

8. Letter from Kenneth C. Lovell to E. L. Bartlett, September 3, 1964, in Record Group 181, Naval Districts and Shore Establishments, 17th Naval District, Alaska: Records of the Office in Charge of Construction, 1944-1977; Naval Petroleum Reserve, no. 4: Historical Files, 1953-74, NARA.

9. John C. Reed and Andreas G. Ronhovde, *Arctic Laboratory: A History, 1947-1966, of the Naval Arctic Research Laboratory at Point Barrow, Alaska* (Washington, DC: Arctic Institute of North America, Office of Naval Research, 1971), 18-19.

10. John C. Reed, "Permafrost and Pet 4," in RG 401, Collection XJCR, John C. Reed Papers, Box 11, Folder: Presentations and Speeches, Subfolder: Permafrost and Pet 4, January 1969, NARA. Robert Legget, "Potentialities of the Northwest: An Engineering Assessment," from Symposium on The Canadian Northwest: Its Potentialities, Royal Society of Canada, 1958, Technical Paper No. 65 of the Division of Building Research, Ottawa, May, 1959, 16-17, found in R. F. Legget Fonds, MG 30 J 44, Vol. 5, File 26, LAC.

11. Quote from "Draft, Statement by the Director of Naval Petroleum and Oil Shale Reserves," RG 181, Naval Districts and Shore Establishments, 17th Naval District, Alaska: Records of the Office in Charge of Construction, 1944-1977: Naval Petroleum Reserve, no. 4: Historical Files, 1953-74, Box 1, Folder: "Historical Matters," NARA-A. John C. Reed, "Pet 4: A Key to Arctic Operations," in RG 401, John C. Reed Papers, Collection XJCR, Box 11, Folder: "Presentations and Speeches," NARA.

12. Letter from John C. Reed to Head, Geography Branch, April 4, 1952, RG 57, Records of John C. Reed, Staff Geologist for Territories and Island Possessions, 1932-53, Box 7, Folder:

Correspondence with Navy, NARA. M. C. Shelesnyak, "The History of the Arctic Research Laboratory, Point Barrow, Alaska," *Arctic* 1, no. 2 (1948): 101. "Arctic Symposium Underway at UA," *Fairbanks Daily News-Miner*, April 10, 1969. See also Reed and Ronhovde, *Arctic Laboratory*, 30–38; and Letter from W. P. Platt to John C. Reed, June 14, 1948, RG 57, Records of the Alaska Terrain and Permafrost Section, 1945–54, Box 14, Folder: John C. Reed (Correspondence), NARA.

13. Reed and Ronhovde, *Arctic Laboratory*, 1–2.

14. Letter from Director, USGS to Rear Admiral Bolster, August 9, 1951, RG 57, Entry: Geologic Division, Records of the Alaska Terrain and Permafrost Section, 1945–54, Box 2, Folder: Arctic Research Laboratory, Data on Barrow, Alaska, NARA; Edmund A. Wright, *CRREL's First 25 Years, 1961–1986* (US Army Cold Regions Research and Engineering Laboratory, June 1986), ii. Scientists within this program also shared data and research methods with the Snow, Ice, and Permafrost Research Establishment of the US Army Corps of Engineers, created in the same year. See also Letter from Frank C. Whitmore to Colonel F. B. Hall, March 2, 1953, RG 57, Records of John C. Reed, Staff Geologist for Territories and Island Possessions, 1932–53, Military Geology Branch: General Correspondence Files, 1943–53, Box 1, Folder: Alaska, 1947–August 1950, Permafrost Activities: Defense Dept., NARA.

15. "Disposition of thermistor cables, Pt. Barrow, Alaska: Permafrost Project," Naval Arctic Research Laboratory Records, "Arctic Research Collection," Box 5, Folder: "Investigations (USGS) Permafrost, 1949–1951," APR-UAF; Dr. R. G. MacCarthy, "Something about Permafrost" (Seminar held at the Naval Arctic Research Laboratory, July 28, 1949), Box 6, Folder: "Seminars," Naval Arctic Research Laboratory Records, "Arctic Research Collection," APR-UAF. On the soil moisture test method, see "Engineers Find Way to Measure Soil Moisture," American Association for Advancement of Science, Alaska Division, 1959–1969 Records, Box 5, Folder 86, APR-UAF. On the test method's connection to finding soil types, see Letter from James A. Roy to Gerald R. McCarthy, November 10, 1951, RG 57, Records of John C. Reed, Staff Geologist for Territories and Island Possessions, 1932–53, Box 2, Geologic Division Files, Alaska Terrain Permafrost Program, NARA. See also "Arctic Research Laboratory Newsletter, Point Barrow, Alaska," no. 3.5, April 1952, RG 57, Records of the Alaska Terrain and Permafrost Section, 1945–54, Box 2, Folder: Arctic Research Laboratory, Data on Barrow, Alaska, NARA. Institutional affiliation given in Letter from Ira L. Wiggins to Dr. G. W. Rowley, March 30, 1951, Naval Arctic Research Laboratory Records, "Arctic Research Collection," Box 5, Folder: "Investigations (USGS) Permafrost, 1949–1951," APR-UAF.

16. Letter from Robert Black to Harry E. Balvin, January 30, 1951, Naval Arctic Research Laboratory Records, Box 5, Folder: "Investigations (USGS) Permafrost, 1949–1951," APR-UAF. Memo from R. F. Black to L. L. Ray, November 23, 1949, RG 57, Geologic Division, Records of the Alaska Terrain and Permafrost Section, 1945–54, Box 4, Folder: Robert Black, NARA.

17. Memorandum for the Files, Subject: Special Meeting Held at 0830, December 4, 1950, in Room 1515, T-3 Building, Navy Department, on the Subject of Production in and through Permafrost, Navy Department, Washington, DC, found in RG 181 (Naval Districts and Shore Establishments, 17th Naval District, Alaska: Records of the Office in Charge of Construction, 1944–77: Naval Petroleum Reserve, no. 4: Historical Files, 1953–74), Box 3: Office of the Officer in Charge of Construction, 1947–51, Reports Relating to Oil Exploration, 1944–53, Folder: "Permafrost Studies," NARA-A. "Progress Report for Arctic Ice-Permafrost Program, 1949–1952, with Projections for Work to be Done in 1953," RG 57, Records of John C. Reed, Staff Geologist for Territories and Island Possessions, 1932–53, Box 2, Geologic Division Files, Alaska Terrain Permafrost Program, NARA. *Anchorage Daily News*, "Max Brewer: Obituary," September 30, 2012, http://www.legacy.com/obituaries/adn/obituary.aspx?pid=160163549, accessed March 13, 2013. See also *Anchorage Daily News*, "Max Brewer: Obituary."

18. Letter from Thomas G. Roberts to Louis L. Ray, March 28, 1951, RG 57, Geologic Division, Records of the Alaska Terrain and Permafrost Section, 1945–54, Box 2, Folder: Arctic Research Laboratory, Data on Barrow, Alaska, NARA. On Reed's list being applied, see Letter from Kenneth C. Lovell to E. L. Bartlett, September 3, 1964, NARA.

19. Reed, "Pet 4: A Key to Arctic Operations." See also Letter from G. R. Rowley to Dr. I. L. Wiggins, March 9, 1951, Naval Arctic Research Laboratory Records, Box 5, Folder: "Investigations (USGS) Permafrost, 1949–1951," APR-UAF.

20. Legget, "Potentialities of the Northwest," 19. Memo to Commodore W. G. Greenman, February 25, 1949, RG 57, Geologic Division, Records of the Alaska Terrain and Permafrost Section, 1945–54, Box 10, Folder: Naval Petroleum Reserves, NARA. Letter from E. L. Davis to Officer in Charge, February 13, 1951, Naval Arctic Research Laboratory Records, "Arctic Research Collection," Box 5, Folder: "Investigations (USGS) Permafrost, 1949–1951," APR-UAF. On estimates of oil and gas, see Reed, "Pet 4: A Key to Arctic Operations," and B. G. Sivertz, "The North as a Region," *Resources for Tomorrow, Volume I, Section Four* (Ottawa: Queen's Printer, 1961), 561. On final quote, see Iris Warner, "The Inuvik Research Laboratory," ARI.

21. "Manuscript Report," 2–3, in RG 401, 46: Alton A. Lindsey Papers, Box 7, Entry 8: Papers Relating to the Purdue-Canadian Arctic Permafrost Expedition, 1951–52, Folder: "Expedition 1951: Manuscript Report," NARA. See also Folder: Alcan Highway Photos and Miscellaneous Maps, RG 401, Paul C. Dalrymple Papers, Box 21, NARA. This folder shows the images and studies commissioned by the Quartermaster corps in the 1940s to better understand a set of Alaskan highways (Glenn, Steese, Richardson, Tok, and Alcan).

22. "Manuscript Report," 3–7, in RG 401, 46: Alton A. Lindsey Papers, Box 7, Entry 8: Papers Relating to the Purdue-Canadian Arctic Permafrost Expedition, 1951–52, Folder: "Expedition 1951: Manuscript Report," NARA. Lindsey corresponded with Porsild after the completion of the permafrost expedition. Letter from Alton A. Lindsey to A. E. Porsild, March 19, 1952, RG 401, 46, Alton A. Lindsey Papers, Box 7, Entry 8: Papers Relating to the Purdue-Canadian Arctic Permafrost Expedition, 1951–52, Folder: "Expedition 1951: Correspondence, 1951–2," NARA.

23. A. L. Washburn, "Classification of Patterned Ground and a Review of Suggested Origins," *Bulletin of the Geological Society of America* 67 (July 1956): 831–33.

24. Warner, "Inuvik Research Laboratory." See also "Manuscript Report," 9. On relations of Imperial Oil and Northern Research Station, see Memo from R. F. Legget to Mr. J. MacMillan, Superintendent of Imperial Oil at Norman Wells, April 28, 1955, RG 85, File 1010-1-1, LAC. On Pihlainen's use of transparencies, see John A. Pihlainen to Mr. J. J. McCarthy, January 27, 1964, RG 77. For Pihlainen's contributions to the field of permafrost, see Roger J. E. Brown, "Obituary: John Aito Pihlainen (1926–1964)," *Arctic* (1964): 142.

25. Letter from Acting Director to Col. W. C. Hall, March 1, 1943, RG 57, Records of John C. Reed, Staff Geologist for Territories and Island Possessions, 1932–53, Military Geology Branch: General Correspondence Files, 1943–53, Box 1, Folder: "Alaska, 1943–July 1945, Permafrost: Vol. I-A," NARA. According to one account, the amount of gravel required for northern airstrips in the 1940s and 1950s was equivalent to laying "a two-lane highway with 12-inch thick roadbed from San Francisco to New York City." See Coates, *Trans-Alaska Pipeline Controversy,* 76–77. On perfection of the aerial photography method, see Letter from H. P. Wilson to Director of Air Services, Controller of Meteorology, September 24, 1951, T-14212, Vol. 1039, File 21525 ("Purdue University/R. E. Frost Permafrost"), LAC.

26. R. F. Legget, "Permafrost Research in Northern Canada," *Nature* (October 1956), found in R. F. Legget Fonds, MG 30 J 44, Vol. 5, File 21, LAC.

27. Dick Hill, *Inuvik: A History, 1958–2008; The Planning, Construction, and Growth of an Arctic Community* (Victoria, BC: Trafford Publishing, 2008), 27. See also Irene Baird, "Inuvik: Place of Man," *Beaver* (1960); "'Gamble on Stilts' Portent of New North," *Red Deer Advocate,* July 27, 1961; and C. L. Merrill, J. A. Pihlainen, and R. F. Legget, "The New Aklavik: Search for the Site," *Engineering Journal* (January 1960): 52–57.

28. Merill, Pihlainen, and Legget, "New Aklavik," 52–54.

29. Ibid., 57.

30. Ibid.

31. Quoted in Dick Hill, "Diefenbaker's 'Opening of Inuvik' Speech," http://www.inuvik heritage.com/Feature1.html, accessed March 13, 2013.

32. "To the President: The Commission appointed by you to consider the Naval Reserve Oil situation respectfully submits the following report of its operations to date," n.d., RG 57, Program and Office Files, Records Concerning Naval Oil Reserves, 1921–27, Box 1, Unmarked Folder, NARA. See also Reed, "Pet 4: A Key to Arctic Operations," 8–13.

33. R. G. Robertson, "The Future of the North," a speech given before Carleton University, 1957, "Northern Development" folder, DHC, 6. See also R. G. Robertson, *Memoirs of a Very Civil Servant: Mackenzie King to Pierre Trudeau* (Toronto: University of Toronto Press, 2000), 146–99.

34. Ernest Gruening, *The State of Alaska* (New York: Random House, 1954), 409–56, 527; and Robertson, "Future of the North," 12–13. This rhetoric of a "will to improve" is common to a colonial "development" discourse after World War II that transcended the Arctic. Tania Murray Li, *The Will to Improve: Governmentality, Development, and the Practice of Politics* (Durham, NC: Duke University Press, 2007).

35. *Imperial Oil Review*, October 1960, found in DHC, Folder: Northern Development. See also Ernest Gruening, ed., *An Alaskan Reader, 1867–1967* (New York: Meredith Press, 1966).

36. "Alaska Petroleum Reserve Data to be made Available by Navy," Department of Defense Office of Public Information Immediate Release, September 14, 1954, found in Naval Arctic Research Laboratory Records, "Arctic Research Collection," Box 7, Folder: "USGS—1955," APR-UAF. See also "Testimony of Mr. Bartlett to Congress, April 28, 1967, 'Alaska Oil Progress Report: Naval Petroleum Reserve no. 4,'" RG 181, Naval Districts and Shore Establishments, 17th Naval District, Alaska: Records of the Office in Charge of Construction, 1944–1977: Naval Petroleum Reserve, no. 4: Historical Files, 1953–1974, Box 12 Folder: "Historical Matters, histories, etc." NARA-A; and B. J. Galloway, "Appendix D: Historical Overview of North Slope Petroleum Development and *Exxon Valdez* Oil Spill," in *Environmental Report for Trans Alaska Pipeline System Right-of-Way Renewal* (February 2001), D-1.

37. "Proceedings: National Northern Development Conference: The Last Frontier in North America, Edmonton, Alberta, Sept. 17, 18, 19, 1958," 38–43, 140–45, and 164.

38. Charles Arnold, Wendy Stephenson, Bob Simpson, and Zoe Ho, eds., *Taimani: At That Time; Inuvialuit Timeline Visual Guide* (Inuvik, NT: Inuvialuit Regional Corporation, 2011), 117.

39. "The Northern Plague" poem was found in *The Inuvik Drum*, vol. 1, no. 23, June 9, 1966, 6. It was written by A. G. "Pat" Kelly, Fort Norman, NT, on June 9, 1959.

40. Merill, Pihlainen, and Legget, "New Aklavik," 57. Peggy Martin Brizinski, "The Summer Meddler: The Image of the Anthropologist as Tool for Indigenous Formulations of Culture," in *Anthropology, Public Policy, and Native Peoples in Canada*, ed. Noel Dyck and James B. Waldram (Montreal: McGill-Queens University Press, 1993), 150.

41. Hill, *Inuvik: A History*, 35. On the Hudson's Bay Company's decisions, see "Frames of Reference and Organization: Coronation Gulf Area Economic Survey," Memorandum from the Department of Northern Affairs and National Resources, May 8, 1963, G 1979 003, Area Economic Survey Assignments Folder, NWTA. On the Family Allowance Act, see *Inuvialuit Pitquisiit: The Culture of the Inuvialuit* (1991), 62. See also David Damas, *Arctic Migrants / Arctic Villagers: The Transformation of Inuit Settlement in the Central Arctic* (Montreal: McGill-Queen's University Press, 2002), 51. Aklavik, Northwest Territories, "History," http://www.aklavik.ca/?p=history, accessed March 13, 2013.

42. Hill, *Inuvik: A History*, 35.

43. For a full treatment on the "modernization theory" as understood through Inuvik, see Matthew Farish, "Frontier Engineering: From the Globe to the Body in the Cold War Arctic," *Canadian Geographer* 50, no. 2 (2006): 177–96. On resentment, see Peggy Martin Brizinski, "The Structure and Image of Social Science Participation in the Mackenzie Delta" (master's thesis, McMaster University, 1982), 46, found in "Anthropology" folder, DHC. Quote about "new governmental scheme" from Brizinski, "Summer Meddler," 150–51.

44. On Inuit employment in Barrow, at Pet-4, at the DEW Line stations nearby, see Reed and Ronhovde, *Arctic Laboratory*, 28–29, as well as Arnold et al., *Taimani*, 114–20. Even after the town's completion, such jobs were readily available through the 1950s, because of the construction of oil wells and the string of radar stations of the Distant Early Warning Line.

CHAPTER 5

1. "Proceedings of the Conference on Productivity and Conservation in Northern Circumpolar Lands, Edmonton, Alberta, 15 to 17 October 1969," 327–29.

2. On this usage of "postcolonial," see Bill Ashcroft, Gareth Griffiths, and Helen Tiffin, *The Empire Writes Back: Theory and Practice in Post-Colonial Literatures* (New York: Routledge, 1989), 2–12. I do not intend this word to suggest some state of the world in which all colonial legacies have been addressed and redressed.

3. On the different uses and politics of "decolonial" and "postcolonial," see Ramon Grosfoguel, "Decolonizing Post-Colonial Studies and Paradigms of Political-Economy: Transmodernity, Decolonial Thinking, and Global Coloniality," *Transmodernity: Journal of Peripheral Cultural Production of the Luso-Hispanic World* 1 (2011), http://escholarship.org/uc/item/21k6t3fq, accessed November 30, 2015.

4. "The Impacts of Oil Activities on the Mackenzie Delta, N.W.T: A Report," prepared for the Department of Indian Affairs and Northern Development, March 1971, published by Information Canada, 1972, 1–19, ARI. See also "A Summary of Oil and Gas Activities North of 60," *Arctic Circular* 22, no. 1 (1972): 49. For Alaska, see "A Historic and Current View of Alaskan Arctic Research and Events," 11, David Hickok Papers, Box 2, Folder 38, UAA. Oil companies like Imperial Oil employed 350 to 400 men for a drilling season on any given year between 1960 and 1970. These men were split into two different kinds of crews—the drilling team and the exploration (or "seismic") team.

5. A. T. Davidson, "Some Factors Regarding Northern Oil and Gas," *Arctic Circular* 12, no. 4 (1960): 56–57.

6. In addition to Prudhoe Bay, Imperial Oil struck a lucrative well at Atkinson Point, and Gulf Canada at Parson's Lake, both in the Mackenzie Delta. The Arctic Islands also boasted some productive wells, though they were distant from the administrative centers built in the northwest corner of the continent over the 1950s. "The Impacts of Oil Activities on the Mackenzie Delta, N.W.T: A Report," sec. 2, p. 2. See also "A Summary of Oil and Gas Activities North of 60," 51–64. On Parson's Lake, see Dick Hill, *Inuvik: A History, 1958–2008; The Planning, Construction, and Growth of an Arctic Community* (Victoria, BC: Trafford Publishing, 2008), 113–14. On Atkinson Point, see Charles Arnold, Wendy Stephenson, Bob Simpson, and Zoe Ho, eds., *Taimani: At That Time; Inuvialuit Timeline Visual Guide* (Inuvik, NT: Inuvialuit Regional Corporation, 2011), 123. On islands, see John C. Reed, "Oil Development and Conservation in Arctic America," *Biological Conservation* 4, no. 5 (1972): 369–70.

7. "The Impacts of Oil Activities on the Mackenzie Delta, N.W.T: A Report," figures 2–4 on pages 2–8.

8. EPEC Consulting Western Ltd., "Impact of Seismic Exploration on Muskrat Populations," prepared for the Arctic Land Use Research Programme, 1976, 1–2, found in ARI. On the advent of the single-line method, see Charles C. Bates, Thomas F. Gaskell, and Robert B. Rice, *Geophysics in the Affairs of Man: A Personalized History of Exploration Geophysics and Its Allied Sciences of Seismology and Oceanography* (New York: Pergamon Press, 1982), 101–15.

9. "Interview of Max Brewer, John Schindler, and Jess Walker by Sam Means, Denise and Deidre Ganopole, and David Hickok, July 18, 1975, Anchorage, Alaska," in David Hickok Papers, Box 35, Folder 31, UAA. See also "Proceedings of the Conference on Productivity and Conservation in Northern Circumpolar Lands," 7.

10. J. Ross Mackay, "Disturbances to the Tundra and Forest Tundra Environment of the Western Arctic," *Canadian Geotechnical Journal* 7, no. 4 (1970): 421.

11. On the lack of resiliency of low productivity, see Stephen Bocking, "Science and Spaces in the Northern Environment," *Environmental History* 12, no. 4 (2007): 867–94. See also Peter A. Coates, *The Trans-Alaska Pipeline Controversy: Technology, Conservation, and the Frontier* (Anchorage: University of Alaska Press, 1993), 98–99. Knowledge of tundra instability came through Lawrence Palmer's work with reindeer over the 1940s. See Lawrence Palmer and Charles H. Rouse, "An Ecological Study of the Alaskan Tundra: With Reference to Its Reactions to Grazing Use," RG 75, Correspondence of Lawrence J. Palmer, 1920–45, Box 2, NARA. On Jerry Brown, see "Proceedings of the Conference on Productivity and Conservation in Northern Circumpolar Lands," 41–47.

12. Typically, historians locate other sites and events in their narratives of environmentalism during the 1960s—namely, Rachel Carson's *Silent Spring* and the Santa Barbara Oil Spill. For a direct discussion of environmentalism in the United States during the 1960s, see Adam Rome, "'Give Earth a Chance': The Environmental Movement and the Sixties," *Journal of American History* 90, no. 2 (2003): 525–54. Dan O'Neill has most clearly articulated the link between the Arctic and modern environmentalism with his treatment of the issue of "fallout" and the ecological study of atomic radiation. See Dan O'Neill, *The Firecracker Boys: H-Bombs, Inupiat Eskimos, and the Roots of the Environmental Movement* (New York: Basic Books, 2007). On ecosystem ecology at midcentury, and its attention to energy flows and mathematical modeling, see Steven Bocking, "Oak Ridge Ecosystem Research and Impact Assessment," chapter 5 in *Ecologists and Environmental Politics: A History of Contemporary Ecology* (New Haven, CT: Yale University Press, 1997), 89–115; and Sharon E. Kingsland, *The Evolution of American Ecology, 1890-2000* (Baltimore: Johns Hopkins University Press, 2008), 206–31.

13. "Proceedings of the Conference on Productivity and Conservation in Northern Circumpolar Lands," 330–31. Bates, Gaskell, and Rice also detail this rift between scientists and paradigms in the 1970s, with attention to geophysicists, acknowledging that not all engineers from the 1950s or 1970s were antienvironmental. See *Geophysics in the Affairs of Man*, 240–45.

14. O'Neill, *Firecracker Boys*. See also "Second Inupiat at Kotzebue," *Tundra Times*, October 1, 1961, 1; and "Encroachments Stressed at Inupiat Paitot Conference," *Tundra Times*, November 1, 1962, 15. See also Maria Shaa Tláa Williams, "A Brief History of Native Solidarity," in *Arctic Voices*, ed. Subhankar Banerjee (New York: Seven Stories Press, 2012), 459.

15. Elizabeth James, "Toward Alaska Native Political Organization: The Origins of *Tundra Times*," *Western Historical Quarterly* 41, no. 3 (2010): 285–303. See also Alexander M. Ervin, "The Emergence of Native Alaskan Political Capacity, 1959-1971," *Muskox* 19 (1976), available online at http://www.alaskool.org/projects/ancsa/ARTICLES/ervin1976/Ervin_MuskOx.htm#THE%20 LEGAL%20ISSUE%200F%20LAND%20RIGHTS%20IN%20ALASKA, accessed June 20, 2015.

16. Hensley's essay directed the activities of Alaska Natives until sovereignty agreements with the US government in 1971. A fellow Inupiaq, Charlie Edwardsen, brought copies of Hensley's paper to the first meeting of the Alaska Native Federation in 1966. William L. Hensley, "What Rights to Land Have the Alaska Natives? The Primary Question," 2001, available at http://www .alaskool.org/projects/ancsa/wlh/wlh66-all-bigtype.pdf, accessed June 20, 2015. See also Paul Ongtooguk, "The Annotated ANCSA," http://www.alaskool.org/projects/ancsa/annancsa.htm, accessed October 30, 2015. On the map biography method of documenting land use and occupancy, see Imre Sutton, ed., *Irredeemable America: The Indians' Estate and Land Claims* (Albuquerque: University of New Mexico Press, 1985).

17. O'Neill, *Firecracker Boys*, see especially chapters 10 to 13.

18. On the composition of Project Chariot's environmental studies committee, and the tensions within it, see O'Neill, *Firecracker Boys*, 162–83, quote from Viereck on 321–22. Barry Commoner, *Science and Survival* (New York: Viking Press, 1963), 117–20, quote from Pruitt on 117–18. Pruitt and Viereck went on to make critiques of oil development in the Arctic based on their biological studies. Foote helped organize a protest at the University of Alaska in 1969—a "research halt"

meant to reveal the misuse of scientific information by technocratic elites in relation to Arctic oil development. He died in a car accident and did not see the event to its completion. O'Neill, *Firecracker Boys*, 286–87.

19. O'Neill, *Firecracker Boys*, 286–87. See also Ervin, "The Emergence of Native Alaskan Political Capacity, 1959–1971." Alaska History and Cultural Studies, "Modern Alaska: Alaska Native Claims Settlement Act," http://www.akhistorycourse.org/articles/article.php?artID=139, accessed June 20, 2015. "Native Groups to Meet: Leaders to Gather Together to Talk," *Tundra Times*, June 8, 1964, 1. Bill Hess, "Taking Control—The Story of Self Determination in the Arctic," http://www.alaskool.org/native_ed/curriculum/aamodt/tc.html, accessed June 20, 2015.

20. In 1959, the first judge of the Territorial Court of the Northwest Territories, John Howard Sissons, decided a landmark case that established the legal basis for Inuit claims to land title. As in Alaska and its 1884 Organic Act, the Canadian record of law used the word "disturbed" to delineate indigenous lands and their relationship to the Crown. In his decision, Sisson quoted the Royal Proclamation of 1763, which acted, essentially, as Inuit treaty, since there was no other treaty or act that applied to "Eskimos." Sisson wrote, "The lands of the Eskimos are reserved to them as their hunting grounds. It is the Royal will that the Eskimos 'should not be molested or disturbed in the possession of these lands.' Others should tread softly, for this is dedicated ground." Quoted in Peter Kulchyski and Frank Tester, *Kiumajut (Talking Back): Game Management and Inuit Rights, 1900–1970* (Vancouver: University of British Columbia Press, 2007), 181. On land use and occupancy studies, see Milton M. R. Freeman, "Looking Back—and Looking Ahead—35 Years after the Inuit Land Use and Occupancy Project," *Canadian Geographer* 55, no. 1 (2011): 20–31. See Nuligak, *I, Nuligak*, ed. and trans. Maurice Metayer (Toronto: Peter Martin, 1966), especially page 157.

21. "Mackenzie Institute Activities and Operations 1968," DHPC.

22. Rose Mary Thrasher, "Petroleum Activities in the Delta," *Delta Newsletter* 1 (1970), DHPC. On relationships among Inuit and biologists in this period, see Norman B. Snow, interview by Andrew Stuhl, Joint Secretariat Office, Inuvik, May 28, 2014.

23. Peggy Martin Brizinski, "The Structure and Image of Social Science Participation in the Mackenzie Delta" (master's thesis, McMaster University, 1982), 20–25. Frank Tester and Peter Kulchyski, *Tammarniit (Mistakes): Inuit Relocation in the Eastern Arctic, 1939–1963* (Vancouver: University of British Columbia Press, 1994), 326–27. Kulchyski and Tester, *Kiumajut*, 186–87. See also David Damas, *Arctic Migrants / Arctic Villagers: The Transformation of Inuit Settlement in the Central Arctic* (Montreal: McGill-Queen's University Press, 2002), 127–29.

24. Peter J. Usher and Hugh Brody, "Obituary: Arnold James (Moose) Kerr (1921–2008)," *Arctic* 63, no. 1 (2010): 121–23.

25. Ibid. Richard M. Hill, e-mail message to the author, February 10, 2012. Tester and Kulchyski acknowledge the parallel existence of old and new guards within the northern administration in the early 1960s. In *Kiumajut*, they write: "While in the 1950s policy makers could operate largely in a vacuum, developing policies for the most part free of criticisms and without strong alternatives, by the early 1960s, the notion of Inuit having Aboriginal rights was underwriting the possibility of an entirely different policy trajectory and implicitly criticizing the direction taken by northern administrators" (191). See also Tina Loo's discussion of community development in Rankin Inlet: Tina Loo, "Hope in the Barrenlands: Northern Development and Sustainability's Canadian History," in *Ice Blink: Navigating Northern Environmental History*, ed. Brad Martin and Stephen Bocking (Calgary: University of Calgary Press, in press). Special thanks to the author for allowing me to review this piece while the volume was in press.

26. R. M. Hill, "New Northern Education Institute," *Arctic Circular* 19, no. 2 (1969): 40. "Competing Discourses on Development," http://arcticcircle.uconn.edu/SEEJ/Landclaims/ancsa4.html, accessed June 20, 2015.

27. Arnold et al., *Taimani*, 133–37. While COPE initially opened its doors to First Nations as well as Métis residents of the entire Northwest Territories, it eventually separated from these groups, seeking by 1972 to represent only the indigenous people of the western Canadian Arctic.

28. Ibid. See also Jean Chretien, "Northern Canada in the 70's," a report to the Standing Committee on Indian Affairs and Northern Development on the Government's northern objectives, priorities and strategies for the 70's, DHC. See also Trevor Lloyd, "Canada's Arctic in the Age of Ecology," *Foreign Affairs* (July 1970). See also "Report of the Mackenzie Delta Task Force, Part II" (1970), Mackenzie Delta Folder, DHC.

29. Telegram from Coppermine Conference to Prime Minister Pierre E. Trudeau, DHC.

30. J. Brooks Flippen, "Richard Nixon and the Triumph of Environmentalism," in *American Environmental History*, ed. Louis Warren (Malden, MA: Blackwell Publishing, 2003), 285–88. Claudia Phillips, "An Evaluation of Ecosystem Management and Its Applications to the National Environmental Policy Act: The Case of the U.S. Forest Service" (PhD diss., Virginia Tech University, 1997), 12–19.

31. Robert B. Gibson, "From Wreck Cove to Voisey's Bay: The Evolution of Federal Environmental Assessment in Canada," *Impact Assessment and Project Appraisal* 20, no. 3 (2002): 151–59. In 1973, Canada formalized its Federal Environmental Assessment Review Process and its Federal Environmental Assessment Review Office.

32. "Conference for Arctic Planning," sponsored by the Admiral Richard E. Byrd Polar Center with the cooperation of the Arctic Institute of North America and the New England Aquarium (October 14–16, 1970), Northern Development Folder, DHC.

33. Coates, *Trans-Alaska Pipeline Controversy*, 176–77.

34. "A Historic and Current View of Alaskan Arctic Research and Events," David Hickok Papers, Box 2, Folder 38, UAA. See also Coates, *Trans-Alaska Pipeline Controversy*, 177–84.

35. James, "Toward Alaska Native Political Organization," 285–303. 43 USC 1601: The Alaska Native Claims Settlement Act, §1601. See also Tom Thornton, "Alaskan Implementation Issues and Related Research Areas," a presentation before Making Treaties Work for Future Generations Conference, December 8, 2015.

36. Coates, *Trans-Alaska Pipeline Controversy*, 179. See also Monica E. Thomas, "The Alaska Native Claims Settlement Act: Conflict and Controversy," *Polar Record* 23 (1986): 27–36.

37. James B. Haynes, "Land Selection and Development under the Alaska Native Claims Settlement Act," *Arctic* 28, no. 3 (1975): 201–8. Thornton, "Alaskan Implementation Issues and Related Research Areas."

38. "Memorandum to Mayor Eben Hopson from Billy Neakok," David Hickok Papers, Box 24, Folder 5, p. 1, UAA.

39. Coates, *Trans-Alaska Pipeline Controversy*, 228–29. "Memorandum to Mayor Eben Hopson from Billy Neakok," UAA.

40. Coates, *Trans-Alaska Pipeline Controversy*, 188, 227. Thomas, "Alaska Native Claims Settlement Act: Conflict and Controversy," 27–36.

41. Coates, *Trans-Alaska Pipeline Controversy*, 262–66. In recent years, critiques of the Alaska Native Claims Settlement Act have become more forceful, especially as the Canadian government continues to negotiate much more comprehensive land claims and self-government agreements with a variety of Aboriginal groups across the provinces and territories. See Tom Thornton, "Alaska Implementation Issues and Related Research Areas," a presentation before the Making Treaties Work for Future Generations: Implementation Research Planning Conference, Ottawa, December 8, 2015.

42. Robert Brent Anderson, *Economic Development among the Aboriginal Peoples of Canada: The Hope for the Future* (North York, ON: Captus Press, 1999), 59–60.

43. Thomas R. Berger, "Northern Frontier, Northern Homeland: The Report of the Mackenzie Valley Pipeline Inquiry," Minister of Supply and Services, 1977, vii–xv.

44. Mark Nuttall summarizes reactions and analyses of the Berger inquiry. See Mark Nuttall, *Pipeline Dreams: People, Environment, and the Arctic Energy Frontier* (Copenhagen: International Work Group for Indigenous Affairs, 2010), 64–72. Importantly, Inuvialuit were not antidevelopment. See Paul Sabin, "Voices from the Hydrocarbon Frontier: Canada's Mackenzie Valley Pipeline Inquiry (1974-1977)," *Environmental History Review* 19, no. 1 (1995): 17–48.

45. Berger, "Northern Frontier, Northern Homeland," 52.

46. "Expanded Guidelines for Northern Pipelines," as tabled in the House of Commons, June 28, 1972, by the honorable Jean Chretien, found in "Alaska Natural Gas Transportation System: Final Environmental Impact Statement," US Department of the Interior (March 1976), available at https://books.google.ca/books?id=teA_AAAAIAAJ&pg=PA559&lpg=PA559&dq=%22 expanded+guidelines+for+northern+pipelines%22&source=bl&ots=xZBjCbgHyS&sig=DkpuCDJLM 2LrQgH2Pp9hSGmT8Dc&hl=en&sa=X&ei=BW6CU9rSBMv20ATY50DoAw#v=onepage &qf=true, accessed June 20, 2015. See also "Report of the Mackenzie Delta Task Force, Part II," 45.

47. Berger, "Northern Frontier, Northern Homeland," vii.

48. "Proposed Land Freezes in the Tuktyaktuk Area," Mackenzie Valley Pipeline Inquiry Exhibit no. 256 put in by Bertram Pokiak, DBC. Canadian Arctic Gas, "Mackenzie Valley Pipeline Inquiry: Phase III, The Impact of a Pipeline and Mackenzie Corridor Development on the Living Environment," ARI, 50-53, 70-72.

49. John David Hamilton, *Arctic Revolution: Social Change in the Northwest Territories, 1935-1994* (Toronto: Dundurn, 1994), 161-166. Arctic Gas's "Environment Protection Board" was comprised of Carson Templeton, who chaired the Alaskan Highway Pipeline Panel; and Max Britton and N. J. Wilimovsky, who sat on the Bioenvironmental Committee of the Atomic Energy Commission's Project Chariot. For composition of this board, see "What about Permafrost," DHC. For links to Project Chariot, see O'Neill, *Firecracker Boys*, 162-65. On reflections on the science of impact assessment, see A. W. F. Banfield, "The Development of Environmental Impact Assessments from a Canadian Perspective," in "Environmental Impact Assessment: A Symposium Sponsored by the Alberta Society of Professional Biologists, Held at Edmonton, Alberta, April 29, 1977," Alberta Society of Professional Biologists, 1977, pp. 74-77. "Mackenzie Valley Pipeline Inquiry: Toronto, Ontario, 27 May 1976: Proceedings at Community Hearings Volume 60" (Allwest Reporting, 2003), 6795.

50. Norman B. Snow, interview by Andrew Stuhl, Joint Secretariat Office, Inuvik, May 28, 2014. On the training of Inuit field-workers, see Arnold et al., *Taimani*, 130-51. On the contents of the Canadian Arctic Gas's reports, see Banfield, "Development of Environmental Impact Assessments from a Canadian Perspective," 75. The sharing of pipeline impacts and close relations among industry, government, and Native organizations seems to be unique to Inuit. On the restriction of information within the Indian Brotherhood, see Earle Gray, *Super Pipe: The Arctic Pipeline—World's Greatest Fiasco?* (Toronto: Griffin House, 1979), 198-200.

51. L. C. Bliss, "The Report of the Mackenzie Valley Pipeline Inquiry, Volume One: An Environmental Critique," *Muskox* 21 (1978): 28-33. R. D. Jackimchuk, "Biological Investigations and Public Participation: The Mackenzie Valley Pipeline," in "Biology, Science Policy, and the Public: A Symposium Sponsored by the Alberta Society of Professional Biologists Held at Edmonton, Alberta, April 27 and 28, 1978," Alberta Society of Professional Biologists, 1978, pp. 73-89. Alaskan scientists had made these points too, especially in wake of approval of pipeline. See Coates, *Trans-Alaska Pipeline Controversy*, 200-205.

52. Quote on "cloistered" biology from Jackimchuk, "Biological Investigations and Public Participation," 82. On the move from private to public, and the creation of a "postindustrial" science, see T. Owen, "Environmental Impact Analysis in Perspective," in *Environmental Impact Assessment in Canada: Processes and Approaches*, ed. M. Plewes and J. B. R. Whitney (Toronto: Institute for Environmental Studies, 1977), 7-14. On the question of acceptable impacts, see Banfield, "Development of Environmental Impact Assessments from a Canadian Perspective," 81-82. These elements of science within impact assessment processes have also been the subject of sociological study. See Brian Campbell, "Uncertainty as Symbolic Action in Disputes among Experts," *Social Studies of Science* 15, no. 3 (1985): 429-53; and Matthew Cashmore, "The Role of Science in Environmental Impact Assessment: Process and Procedure versus Purpose in the Development of Theory," *Environmental Impact Assessment Review* 24 (2004): 403-26.

53. Arnold et al., *Taimani*, 140–60.

54. Hess, "Taking Control." See also Eben Hopson, "Testimony before the Berger Inquiry on the Experience of the Arctic Slope Inupiat with Oil and Gas Development in the Arctic, 1976," Eben Hopson Archives, http://www.ebenhopson.com/papers/1976/BergerSpeech.html, accessed June 20, 2015. "Financial Performance of Native Regional Corporations," *Alaska Review of Social and Economic Conditions* 28, no. 2 (1991): 1–22.

55. "Hopson Criticizes EIS Statements before Gas Pipeline CEQ Hearings," *Arctic Coastal Zone Management Program* 4 (May–June 1977). See also Hess, "Taking Control."

56. Eben Hopson, "Science in the Arctic," *Arctic Coastal Zone Management Program* 7 (October 1977). On hiring of Bob Delury, see Norman B. Snow, interview by Andrew Stuhl, Joint Secretariat Office, Inuvik, May 28, 2014. Arnold et al., *Taimani*, 140–60. On the ICC, see ICC Alaska, "ICC's Beginning," http://www.iccalaska.org/servlet/content/the_beginning.html, accessed November 19, 2015.

57. On negotiations, see J. G. Nelson, "Volume 11: Summary and Recommendations," Inuit Tapirisat of Canada Renewable Resources Project (1975), ARI. Joint Secretariat, "History of the Joint Secretariat," http://www.jointsecretariat.ca/history.html, accessed October 29, 2015. Initially, comanagement boards could only advise federal agencies on Arctic affairs, but by the mid-1980s, they became decision-making bodies—such that no license for resource development can be obtained without their approval. The scientist first appointed by Inuvialuit to chair these boards was Carson Templeton—who was chairman of the Environment Protection Board during the Mackenzie Valley Pipeline Inquiry and the Chairman of the Alaska Highway Pipeline Panel. "Carson Templeton," *Winnipeg Free Press*, http://passages.winnipegfreepress.com/passage-details/id-89334/name-Carson_Templeton_/, accessed November 12, 2015. Gail Osherenko, "Eye-Opening Journey through the Villages of Alaska," *Christian Science Monitor*, January 22, 1986.

58. *Remembrances on the Event of the 50th Anniversary of the NARL, 1947–1997*, Naval Arctic Research Laboratory Collection—NARL 50th Anniversary Remembrance, Box 1 of 1, folder 1, p. 2. "Robert Rausch Selected for Arctic Science Prize," *Arctic Policy Review* (July–August 1984): 2. On research licensing, see Assembly of Northwest Territories, 11th Assembly, Second Session, Friday, February 19, 1988. See also Brizinski, "Structure and Image of Social Science Participation in the Mackenzie Delta," 25–32.

59. On interpretations of native environmental bureaucracies as neocolonial, see Paul Nadasdy, "The Politics of TEK: Power and the Integration of Knowledge," *Arctic Anthropology* 36, nos. 1–2 (1999): 1–18; Arthur Mason, "The Rise of an Alaskan Native Bourgeoisie" *Etudes/Inuit/Studies* 26, no. 2 (2002): 5–22; Mark Nuttall, "Epistemological Conflicts and Cooperation in the Circumpolar North," in *Globalization and the Circumpolar North*, ed. Lassi Heininen and Chris Southcott (Fairbanks: University of Alaska Press, 2010), 160. Nuttall also notes that "modern treaties" emphasized resource governance over land sales. Nuttall, *Pipeline Dreams*, 81. Although I focus exclusively on comanagement boards and impact review bodies, it is important to point out, as Emilie Cameron has, that land claims agreements meant much more to Inuit than a means of decision making. Cameron writes, Inuit sought to "make decisions in the Inuit way" and to design institutions based on the values of Inuit society. Emilie Cameron, *Far Off Metal River: Inuit Lands, Settler Stories, and the Making of the Contemporary Arctic* (Vancouver: University of British Columbia Press, 2015), 181–82. See also Terry Fenge and Paul Quassa, "Negotiating and Implementing the Nunavut Land Claims Agreement," *Options Politiques* (July–August 2009).

60. Walter and Duncan Gordon Foundation, *Rethinking the Top of the World: Arctic Public Opinion Survey, Volume 2* (Fall 2015): 7–8, http://gordonfoundation.ca/sites/default/files/publications/APO_Survey_Volume%202_0.pdf, accessed October 1, 2015. These concerns began in the 1980s and have persisted. See Brizinski, "Structure and Image of Social Science Participation in the Mackenzie Delta," 30–40.

61. Darrell Christie, the coordinator of the Environmental Impact Screening Committee of the Joint Secretariat, expressed these sentiments to me during a conversation on May 23, 2014. See also Bob Simpson, "The Inuvialuit Claim: A Case Study," a presentation before the Making Treaties Work for Future Generations Conference, December 8, 2015; and Graham White, "Research on Co-Management Boards," a presentation before the Making Treaties Work for Future Generations Conference, December 8, 2015.

EPILOGUE

1. Epigraph from Vilhjalmur Stefansson, "Arctic Headlines in Fact and Fable," *Foreign Policy Association* 51 (January 1945): 5.

2. Walter and Duncan Gordon Foundation, *Rethinking the Top of the World: Arctic Public Opinion Survey, Volume 2* (Fall 2015), http://gordonfoundation.ca/sites/default/files/publications/APO _Survey_Volume%202_0.pdf, accessed October 1, 2015. See especially 5–6. Northern Canadians living in the Northwest Territories perceived southern conceptions of the Arctic as not reflecting the positive or negative realities of life on the ground.

3. John English, *Ice and Water: Politics, Peoples, and the Arctic Council* (Toronto: Allen Lane, 2013). Lloyd Axworthy, "A Brief History of the Creation of the Arctic Council," Walter and Duncan Gordon Foundation, http://gordonfoundation.ca/sites/default/files/publications/Axworthy _2010-12-02_ArcticCouncilHistory_Summary.pdf, accessed May 29, 2015. See also Lassi Heininen, "Circumpolar International Relations and Cooperation," in *Globalization and the Circumpolar North*, ed. Lassi Heininen and Chris Southcott (Fairbanks: University of Alaska Press, 2010), 280.

4. Arctic Climate Impact Assessment, *Scientific Report* (Cambridge: Cambridge University Press, 2005). Barrow Declaration, on the Occasion of the Second Ministerial Meeting of the Arctic Council, 2000. On the role of technical impact assessments as means of securing indigenous sovereignty and generating science relevant to policy, see Terry Fenge, "The Arctic Council: Past, Present, and Future Prospects with Canada in the Chair from 2013 to 2015," *Northern Review* 37 (Fall 2013): 7–35.

5. Fenge, "Arctic Council." Nina Wormbs has suggested that the scope and language of technical assessments may privilege natural sciences over social sciences and humanities, which, in turn, can present problems in how northern communities are framed in final reports. See Nina Wormbs, "The Assessed Arctic: How Monitoring Can Be Silently Normative," in *The New Arctic*, ed. Birgitta Evengård, Joan Nymand Larsen, and Øyvind Paasche (Cham, Switzerland: Springer, 2015), 291–302. By 2011, the Arctic Climate Impact Assessment had 3,700 citations in the scientific and technical literature. David P. Stone, *The Changing Arctic Environment: The Arctic Messenger* (New York: Cambridge University Press, 2015), 208.

6. Quoted in Richard Ellis, *On Thin Ice: The Changing World of the Polar Bear* (New York: Vintage, 2010), 333–34. "Former Inuit Circumpolar Leader Sheila Watt-Cloutier Just Misses Nobel Peace Award for Her Work on Climate Change," Inuit Circumpolar Council Canada, October 12, 2007, http://www.inuitcircumpolar.com/former-inuit-circumpolar-leader-sheila-watt-cloutier-just-misses-nobel-peace-award-for-her-work-on-climate-change.html, accessed October 1, 2015.

7. This summary of Gore, the Inuit Circumpolar Council, and the Arctic Council is drawn from English, *Ice and Water*, 154–281. The quote about romanticization is from 216. Inuvialuit Rosemarie Kuptana was a key negotiator in the council's early years. For a perspective on the science of the Arctic Council's impact assessment reports, and how they might lead to disengagement with human dimensions of the region, see Stone, *Changing Arctic Environment*, 25–37, 70–71. On Greenpeace's policies of banning, and their relation to a polar framework, see Anthony Speca, "Let's Ban Bans in the Arctic," *Northern Public Affairs*, July 27, 2012, http://www.northernpublic affairs.ca/index/lets-ban-bans-in-the-arctic/, accessed December 9, 2015.

8. English, *Ice and Water*, 208–9. On the "negligible" comment, see ibid., 267.

9. Becky Rynor, "Indigenous Voices 'Marginalized' at Arctic Council: Inuit Leaders," http://www.ipolitics.ca/2011/11/07/indigenous-voices-marginalized-at-arctic-council-inuit-leaders/, accessed November 15, 2012. The hurdle of financial and human capacity is common among indigenous peoples of the Arctic. See Gail Fondahl, Viktoriya Filippova, and Liza Mack, "Indigenous Peoples in the New Arctic," in Evengård, Larsen, and Paasche, *New Arctic*, 13.

10. Kitigaaryuit Declaration, Inuit Circumpolar Council Canada, http://www.inuit circumpolar.com/declaration---2014.html, accessed October 1, 2015. "Inuit Circumpolar Council: COP21 Position Paper," *Arctic Portal: The Arctic Gateway*, December 3, 2015, http://arcticportal .org/library/news/1633-inuit-circumpolar-council-cop21-position-paper, accessed December 10, 2015. US Department of State, "Conference on Global Leadership in the Arctic: August 30–31, 2015," http://www.state.gov/e/oes/glacier/index.htm, accessed October 1, 2015. In 2011, the council completed another assessment, *Climate Change and the Cryosphere: Snow, Water, Ice, and Permafrost in the Arctic*, which encouraged ministers of Arctic nations to formally recognize greenhouse gas mitigation as the "backbone" of meaningful climate change policy. Stone, *Changing Arctic Environment*, 209.

11. Stone, *Changing Arctic Environment*, 267–77.

12. Peter J. Usher and Hugh Brody, "Obituary: Arnold James (Moose) Kerr (1921–2008)," *Arctic* 63, no. 1 (2010): 121–23. Norman B. Snow, interview by Andrew Stuhl, Joint Secretariat Office, Inuvik, May 28, 2014. See also Alaska Native Science Commission, "Current Projects," http://www.nativescience.org/html/current_projects.html, accessed November 12, 2015.

13. C. R. Burn, "J. Ross Mackay (1915–2014)," *Arctic* 68, no. 1 (2015): 129–31. See also the work of climate change researchers like Bernie Zak at Barrow and Terry Chapin from the University of Alaska-Fairbanks from the late 1900s to the early 2000s, in Charles Wohlforth, *The Whale and the Supercomputer: On the Northern Front of Climate Change* (New York: North Point Press, 2005), especially chapter 6 and chapter 8.

14. See the management of community-scientist partnerships in climate change research through the National Science Foundation, the Barrow Arctic Science Consortium, and the Utqiagvik Inupiat Corporation. "Ice Stories: Dispatches from Polar Scientists: Barrow, Alaska," http://icestories.exploratorium.edu/dispatches/big-ideas/barrow-alaska/, accessed November 12, 2015.

15. Fae L. Korsmo and Amanda Graham, "Research in the North American North: Action and Reaction," *Arctic* 55, no. 4 (2002): 319–28. On Arctic climate modelers who have never been to the Arctic, and their difficulties or inability to incorporate human concerns and parameters, see Wohlforth, *Whale and the Supercomputer*, 141–70 and 273–74. Wohlforth also recounts the ways the National Science Foundation sets the terms of grant programs for Arctic climate research—and thus scientific practice and discourse—by both responding to and creating thematic areas. These can sometimes become academic "fads" and lip service, missing opportunities to create real relationships between residents and researchers or direct research to local concerns. See ibid., 180–99. Geographer Emilie Cameron recently surveyed literature on the human dimensions of climate change and found that social scientists ignore colonialism's impacts on modern Inuit society and the Arctic environment. Cameron, "Securing Indigenous Politics: A Critique of the Vulnerability and Adaptation Approach to the Human Dimensions of Climate Change in the Canadian Arctic," *Global Environmental Change* 22 (2012).

16. On models of research arrangements, see Korsmo and Graham, "Research in the North American North," 322. On environmental monitoring in the Northwest Territories and Inuvialuit Settlement Region, see Environment and Natural Resources "Cumulative Impact Monitoring (NWT CIMP): About Us," http://www.enr.gov.nt.ca/programs/nwt-cimp/about-us, accessed November 2, 2015. These monitoring activities are constitutional obligations of the comprehensive land claims that came after the Inuvialuit Final Agreement in 1984. For an academic level partnership, see Gary Kofinas with Old Crow, Aklavik, Fort McPherson, and Arctic Village, "Community

Contributions to Ecological Monitoring: Knowledge Co-Production in the US-Canada Arctic Borderlands," in *Frontiers in Polar Science: Indigenous Observations of Environmental Change*, ed. Igor Krupnik and Dyanna Jolly (Fairbanks, AK: ARCUS, 2002): 54–92.

17. Andrew Pleasant, "Letters: The Risks and Advantages of Framing Science," *Science* 317 (August 2007): 1168. See also Tania Lombrozo, "How Psychology Can Save the World from Climate Change," National Public Radio, http://www.npr.org/sections/13.7/2015/11/30/457835780/how -psychology-can-save-the-world-from-climate-change, accessed December 1, 2015.

18. Office of the Press Secretary, *Interagency Working Group on Coordination of Domestic Energy Development and Permitting in Alaska*, Executive Order 13580, https://www.whitehouse.gov /the-press-office/2011/07/12/executive-order-13580-interagency-working-group-coordination -domestic-en; Office of the Press Secretary, *Improving Performance of Federal Permitting and Review of Infrastructure Projects*, Executive Order 13604, https://www.whitehouse.gov/the-press -office/2012/03/22/executive-order-improving-performance-federal-permitting-and-review -infr. On some of the impacts of this latter executive order on participatory democracy, see Joan Brunwasser, "Obama's Keystone XL Pipeline Veto Just Smoke and Mirrors?" *OpEdNews.Com*, www.opednews.com/articles/Obama-s-Keystone-XL-Veto-J-by-Joan-Brunwasser-Activism-Environ mental_Climate_Environment-Policy_Environment-Ecology-151208-692.ht, accessed December 8, 2015.

19. Northern Alaska Environmental Center, "Feds Rush Incomplete Environmental Review of Oil and Gas Drilling in Arctic Ocean," http://northern.org/feds-rush-incomplete-environmental -review-of-oil-and-gas-drilling-in-arctic-ocean, accessed June 23, 2015; Yereth Rosen, "Interior Agencies Release Proposed New Arctic Ocean Drilling Standards," *Alaska Dispatch News*, February 20, 2015, http://www.adn.com/article/20150220/interior-agencies-release-proposed-new-arctic -ocean-drilling-standards, accessed October 14, 2015. Pembina Institute, "Joint Review Panel says full suite of Mackenzie recommendations needed for responsible Northern development," http://www.pembina.org/media-release/1988, accessed June 23, 2015; Canadian Press, "Ottawa Greenlights Arctic Offshore Tests over Inuit Objections," Canadian Broadcasting Company News, http://www.cbc.ca/news/canada/north/ottawa-greenlights-arctic-offshore-seismic-tests-over -inuit-objections-1.2688040, accessed June 23, 2015. For a detailed look at how the US federal government has inconsistently implemented environmental regulation in Arctic Alaska since the 1990s, with pressure from the oil lobby, see Pamela A. Miller, "Broken Promises: The Reality of Big Oil in America's Arctic," in *Arctic Voices: Resistance at the Tipping Point*, ed. Subhankar Banerjee (New York: Seven Stories Press, 2012), 179–206.

20. Yereth Rosen and Dermot Cole, "Point Hope Opts Out of Lawsuit over 2008 Chukchi Lease Sale," *Alaska Dispatch News*, March 24, 2015, http://www.adn.com/article/20150324/point -hope-opts-out-lawsuit-over-2008-chukchi-lease-sale, accessed October 14, 2015.

21. Dali Carmichael, "ICC Unites over Resource Development in Far North," *Northern Journal*, http://norj.ca/2014/07/icc-unites-over-resource-development-in-far-north/, accessed June 23, 2015; Dan Joling, "Groups Sue Agency to Block Shell's Arctic Offshore Drilling," *Times Union*, http:// www.timesunion.com/news/science/article/Groups-sue-agency-to-block-Shell-s-Arctic-6302 064.php, accessed June 23, 2015.

22. "Court Sets Aside Award Ruling against Ottawa in Nunavut Lawsuit," http://www.cbc .ca/news/canada/north/court-sets-aside-award-ruling-against-ottawa-in-nunavut-lawsuit -1.2624164, accessed December 8, 2015.

23. Canadian Broadcasting Company, "Humane Society Says It Doesn't Oppose Inuit Seal Hunt," http://www.cbc.ca/news/canada/north/humane-society-says-it-doesn-t-oppose-inuit-seal -hunt-1.2603306, accessed June 22, 2015; Eva Holland, "#Sealfie vs. #Selfie, One Year Later," *Pacific Standard*, http://www.psmag.com/nature-and-technology/sealfie-vs-selfie-one-year-later-ellen -degeneres-and-the-clubbing-of-seals, accessed June 22, 2015. Heather Exner-Pirot, "Commentary: #SavetheArctic . . . from Greenpeace," *Nunatsiaq Online*, http://www.nunatsiaqonline.ca/stories /article/65674save_the_arctic.....from_greenpeace/, accessed June 22, 2015.

24. Emilie Cameron argues forcefully that Inuit should not be responsible for challenging outdated notions of the Arctic—the responsibility lies with southerners. See Emilie Cameron, *Far Off Metal River: Inuit Lands, Settler Stories, and the Making of the Contemporary Arctic* (Vancouver: University of British Columbia Press, 2015). Bob Reiss captures the frustrations of Inupiat leader, and North Slope borough mayor, Edward Itta, in Itta's interactions with Shell Oil, environmental groups, the Alaska legislature, north slope Inupiat communities, and climate scientists. Reiss recommends a streamlining of the federal bureaucracy for permitting offshore oil and gas exploration as a means of ensuring the US "empire" in the Arctic and also allowing Inuit to better control their own fate. I am skeptical of this conclusion and interpret frustrations here as a necessary alternative to centralized decision-making structures. Precaution and slowness today represent a hard-earned victory over the military-industrial complex of the 1950s and should not be abandoned in the face of rapid climate changes or shifts in commodity prices. See Bob Reiss, *The Eskimo and the Oil Man: The Battle at the Top of the World for America's Future* (New York: Business Plus, 2012), 166-215, 275-87.

25. Walter and Duncan Gordon Foundation, *Rethinking the Top of the World*, 7-8. These concerns began in the 1980s and have persisted. See Peggy Martin Brizinski, "The Structure and Image of Social Science Participation in the Mackenzie Delta" (master's thesis, McMaster University, 1982), 30-40. For a personal plea from a Point Hope Native on the difficulties of keeping up with governmental and industrial decision-making processes, see Robert Thompson, Rosemary Ahtuangaruak, Caroline Cannon, and Eark Kingik, "We Will Fight to Protect the Arctic Ocean and Our Way of Life," in Banerjee, *Arctic Voices*, 326-27. See also Jim Aldridge, "Implementation of the Environmental Assessment and Protection Chapter of the Nisga'a Treaty (a microcosm of the broader implementation challenges)," a presentation before the Making Treaties Work for Future Generations Conference, December 8, 2015.

26. Sara Jerving, Katie Jennings, Masako Melissa Hirsch, and Susanne Rust, "What Exxon Knew about the Earth's Melting Arctic," *Los Angeles Times*, October 9, 2015. These tactics are similar to those described by other scholars interested in science and doubt in post-World War II controversies. See Naomi Oreskes and Erik M. Conway, *Merchants of Doubt: How a Handful of Scientists Obscured the Truth on Issues from Tobacco Smoke to Global Warming* (New York: Bloomsbury Press, 2010), 1-9. On the fraudulent survey activities, and their effects in Alaskan politics and communities along the Alaskan portion of the Beaufort Sea, see Thompson, Ahtuangaruak, Cannon, and Kingik, "We Will Fight to Protect the Arctic Ocean and Our Way of Life," 302-11.

27. Geographers have called on historians to speak up on climate change, to situate it as a social and ecological phenomenon in the spaces in which its effects are most visible. Michael Bravo, "Voices from the Sea Ice: The Reception of Climate Impact Narratives," *Journal of Historical Geography* 35 (2009): 256-78. "Inuvialuit Regional Corporation Guidelines for Research in the Inuvialuit Settlement Region," http://nwtresearch.com/sites/default/files/inuvialuit-regional -corporation.pdf, accessed December 2, 2015. See also Morgan Moffitt, Courtney Chetwynd, and Zoe Todd, "Interrupting the Northern Research Industry: Why Northern Research Should Be in Northern Hands," *Northern Public Affairs*, http://www.northernpublicaffairs.ca/index/interrupt ing-the-northern-research-industry-why-northern-research-should-be-in-northern-hands/, accessed December 6, 2015. A prime example of a mutually beneficial research project is the Toxic Legacies Project, a collaboration of researchers at Memorial and Lakehead Universities, the Goyatiko Language Society, and Alternatives North. "Welcome to Abandoned Mines in Northern Canada," http://www.abandonedminesnc.com/, accessed December 10, 2015.

28. Christine Shearer, *"From Kivalina: A Climate Change Story,"* in Banerjee, *Arctic Voices*, 207-19. Eben Hopson, "Hopson's Address to the People of Kodiak and the User's Panel, OSC-Environmental Assessment Program," Eben Hopson Archives, http://ebenhopson.com/papers/1976 /KodiakOCS.html, accessed May 29, 2014.

29. On the Mackenzie Valley Pipeline, see Lauren Krugel, "N.W.T. Minister Hopes Mackenzie Gas Project Permit Doesn't Lapse," CBC News, http://www.cbc.ca/news/canada/north/n

-w-t-minister-hopes-mackenzie-gas-project-permit-doesn-t-lapse-1.3066280, accessed October 1, 2015. On Shell Oil, see Mia Bennett, "Mood at Arctic Energy Summit Subdued Following Shell's Withdrawal," Cryopolitics, http://cryopolitics.com/2015/09/29/mood-at-arctic-energy-summit-sub dued-following-shells-withdrawal/, accessed October 1, 2015.

30. Greenpeace, "Save the Arctic," http://www.greenpeace.org/usa/arctic/, accessed October 15, 2015. Quote from Rex Rock in Bennett, "Mood at Arctic Energy Summit Subdued Following Shell's Withdrawal." On connections of oil revenues, scientific research, and other forms of development, see Krista Langlois, "Barrow, a Town Divided over Shell's Drilling," *Alaska Dispatch News*, July 19, 2015, http://www.adn.com/article/20150719/barrow-town-divided-over-shells-drilling, accessed October 15, 2015. For Nellie Cournoyea's letter to Bart Cahir, as well as Imperial Oil's explanation of the influence of fracking and global gas prices on its decisions around Arctic development, see Guy Quenneville, "Groups Asked to Weigh in on Mackenzie Gas Project Deadline Extension," CBC News, http://www.cbc.ca/news/canada/north/groups-asked-to-weigh -in-on-mackenzie-gas-project-deadline-extension-1.3339213, accessed December 2, 2015. The letter is available at https://docs.neb-one.gc.ca/ll-eng/llisapi.dll/fetch/2000/90464/90550/338535/338661 /2855874/2857344/A74023-1_Letter_to_IORVL_-_Mackenzie_Gas_Project_-_Request_for_an _Extension_of_the_Sunset_Clauses_-_Procedural_Notice_-_A4V5G9.pdf?nodeid=2857021&vernum =-2, accessed January 26, 2016.

31. The "struggle to live rightly" is a phrase used by William Cronon in his foreword to the paperback edition, in William Cronon, ed., *Uncommon Ground: Rethinking the Human Place in Nature* (New York: W. W. Norton, 1996), 22.

Selected Bibliography

MAGAZINES, NEWSPAPERS, NEWSLETTERS, AND CIRCULARS

Aklavik Journal
Anchorage Times
Arctic Circular
Beaver
Fairbanks Daily News-Miner
Globe and Mail
Inuvik Drum
London Times
National Geographic
New York Times
San Francisco Examiner
Tundra Times

ACADEMIC JOURNALS

American Anthropologist
American Museum Journal
Arctic
Bulletin of the American Geographical Society
Bulletin of the American Museum of Natural History
Canadian Geotechnical Journal
Geographical Journal
Polar Record

INFORMALLY PUBLISHED SOURCES (PRESENTATIONS, REPORTS, THESES)

Annual Reports of the American Museum of Natural History
Report of the Canadian Arctic Expedition, 1913–18, all volumes
Report on the Progress of Mineral Investigations in Alaska (US Department of Interior), 1900–1920

Adcock, Christina. "Tracing Warm Lines: Northern Canadian Exploration, Knowledge, and Memory, 1905-65." PhD diss., Scott Polar Institute, Cambridge University, 2010.

Bonnycastle, R. H. G. "Canada's Reindeer Experiment." *Proceedings of the North American Wildlife Conference, Feb. 3-7, 1936*, 424-27. Senate Committee Print, 74th Cong., 2nd sess. Washington, DC, 1936.

Bruno, Andy. "Making Nature Modern: Economic Transformation and the Environment in the Soviet North." PhD diss., University of Illinois at Urbana-Champaign, August 2011.

Button, David. "The Role of Fur Trade Technologies in Adult Learning: A Study of Selected Inuvialuit Ancestors at Cape Krusenstern, NWT (Nunavut), Canada, 1935-1947." PhD diss., University of Calgary, 2008.

Cameron, Emilie. "Scaling Climate: Critical Geographies of Arctic Climate Change." A presentation before the Critical Climate Change Workshop, University of Minnesota-Minneapolis, April 5, 2013.

Cournoyea, Nellie. "Adaptation and Resilience: The Inuvialuit Story." A speech before the 18th Inuit Studies Conference, Washington, DC, October 26, 2012.

Dathan, Patricia Wendy. "The Reindeer Years: Contribution of A. Erling Porsild to the Continental Northwest." Master's thesis, Department of Geography, McGill University, 1988.

Graham, Amanda. "The University That Wasn't: The University of Canada North, 1970-1985." Master's thesis, Lakehead University, 1994.

Lyons, Natasha, Kate Hennessy, Charles Arnold, and Mervin Joe. "The Inuvialuit Smithsonian Project." Reports produced for the Smithsonian Institution.

Phillips, Claudia. "An Evaluation of Ecosystem Management and Its Applications to the National Environmental Policy Act: The Case of the U.S. Forest Service." PhD diss., Virginia Tech University, 1997.

"Proceedings of the Conference on Productivity and Conservation in Northern Circumpolar Lands, Edmonton, Alberta, 15 to 17 October 1969," International Union for Conservation of Nature and National Resources Publications, n.s., no. 16, 1970, 327-29.

Robinson, Michael. "Blonde Eskimos and Yellow Journalism: Reforming the Arctic Narrative in Progressive America." A paper presented to the History of Science Society, 1996.

Sandlos, John. "Where the Reindeer and Inuit Should Play: Animal Husbandry and Ecological Imperialism in Canada's North." Unpublished manuscript.

Sawchuck, Christina. "Amateur Explorers and the 'Opening' of the Canadian North." A paper presented at the Canadian Historical Association annual meeting, June 3, 2008.

Serreze, Mark. "The Arctic as the Messenger of Global Climate Change." Presentation delivered before the Inuit Studies Conference, Washington, DC, October 26, 2012.

Simpson, Bob. "Inuvialuit Research Agenda." Presentation before the NWT International Polar Year Results Conference, January 20, 2011.

Sonnenfeld, Joseph. "Changes in Subsistence among the Barrow Eskimo." PhD diss., Johns Hopkins University, 1956.

US Department of the Interior. Hearings of the Reindeer Committee in Washington, DC, February-March 1931. Washington, DC: Office of the Secretary, 1931.

FORMALLY PUBLISHED SOURCES
(BOOKS, ARTICLES, REPORTS)

Abel, Kerry, and Ken Coates. *Northern Visions: New Perspectives on the North in Canadian History.* Toronto: University of Toronto Press, 2001.

Allen, Arthur James. *A Whaler and Trader in the Arctic, 1895 to 1944: My Life with the Bowhead.* Anchorage: Alaska Northwest Publishing, 1978.

Alunik, Ishmael, Eddie Dean Kolausok, and David Morrison. *Across Time and Tundra: The Inuvialuit of the Western Arctic*. Seattle: University of Washington Press, 2003.

Anderson, Alun. *After the Ice: Life, Death, and Geopolitics in the New Arctic*. New York: Smithsonian Books, 2009.

Andrews, C. L. *The Eskimo and His Reindeer in Alaska*. Caldwell, ID: Caxton Printers, 1939.

Arctic Shadows: The Arctic Journeys of Dr. R. M. Anderson. DVD. Directed by David Gray. Mountain Studios and Grayhound Information Services, 2009.

Barrow, Mark. *A Passion for Birds: American Ornithology after Audubon*. Princeton, NJ: Princeton University Press, 2000.

Berger, Thomas R. *Northern Frontier, Northern Homeland: Report of the Mackenzie Valley Pipeline Inquiry*. Vol. 1. Ottawa: Minister of Supply and Services Canada, 1977.

Bethune, W. C. *Canada's Western Northland: Its History, Resources, Population, and Administration*. Department of Mines and Resources, Lands, Parks, and Forests Branch, Ottawa, 1937.

Blackman, Margaret. *Sadie Brower Neakok: An Iñupiaq Woman*. Seattle: University of Washington Press, 1989.

Bloom, Lisa. *Gender on Ice: American Ideologies of Polar Expeditions*. Minneapolis: University of Minnesota Press, 1993.

Bocking, Stephen. "A Disciplined Geography: Aviation, Science, and the Cold War in Northern Canada, 1945-1960." *Technology and Culture* 50, no. 2 (2009): 265-90.

———. "Science and Spaces in the Northern Environment." *Environmental History* 12, no. 4 (2007): 867-94.

———. "Situated yet Mobile: Examining the Environmental History of Arctic Ecological Science." In *New Natures: Joining Environmental History with Science and Technology Studies*, edited by Dolly Jørgensen, Finn Arne Jørgensen, and Sara B. Pritchard, 164-78. Pittsburgh: University of Pittsburgh Press, 2013.

Bockstoce, John R. *Furs and Frontiers in the Far North: The Contest among Native and Foreign Nations for the Bering Strait Fur Trade*. New Haven, CT: Yale University Press, 2009.

———. *Whales, Ice, and Men: The History of Whaling in the Western Arctic*. Seattle: University of Washington Press, 1986.

Bockstoce, John R., and Daniel B. Botkin. "The Historical Status and Reduction of the Western Arctic Bowhead Whale (*Balaena mysticetus*) Population by the Pelagic Whaling Industry, 1848-1914." *Scientific Reports of the International Whaling Commission*. Special Issue, no. 5 (1983): 107-41.

Bodfish, Hartson. *Chasing the Bowhead*. Cambridge, MA: Harvard University Press, 1936.

Bowler, Peter J. *The Fontana History of the Environmental Sciences: Geography, Geology, Oceanography, Meteorology, Natural History, Paleontology, Evolution Theory, Ecology*. London: Fontana, 1992.

———. "The Geography of Extinction: Biogeography and the Expulsion of 'Ape Men' from Human Ancestry in the Early Twentieth Century." In *Ape, Man, ApeMan: Changing Views since 1600*, edited by Raymond Corbey and Bert Theunissen, 185-92. Evaluative Proceedings of the Symposium "Ape, Man, Apeman: Changing Views since 1600," Leiden, Netherlands, 28 June-1 July, 1993, Leiden University.

———. *Life's Splendid Drama: Evolutionary Biology and the Reconstruction of Life's Ancestry, 1860-1940*. Chicago: University of Chicago Press, 1996.

Braun, Bruce. *The Intemperate Rainforest: Nature, Culture, and Power on Canada's West Coast*. Minneapolis: University of Minnesota Press, 2002.

Bravo, Michael. "Ethnographic Navigation and the Geographic Gift." In *Geography and Enlightenment*, edited by David Livingstone and Charles Withers. Chicago: University of Chicago Press, 1999.

———. "Measuring Danes and Eskimos." In *Narrating the Arctic: A Cultural History of Nordic Scientific Practices*, edited by Michael Bravo and Sverker Sörlin. Canton, MA: Science History Publications, 2002.

———. "Voices from the Sea Ice: The Reception of Climate Impact Narratives." *Journal of Historical Geography* 35 (2009): 256–78.

Bravo, Michael, and Sverker Sörlin. *Narrating the Arctic: A Cultural History of Nordic Scientific Practices.* Canton, MA: Science History Publications, 2002.

Brock, Reginald. Preface to *My Life with the Eskimo,* by Vilhjalmur Stefansson. 2nd ed. New York: Macmillan, 1924.

Brooks, Alfred Hulse. "The Physiographic Provinces of Alaska." In *Source Book for the Economic Geography of North America,* edited by Charles C. Coby, 412–13. Chicago: University of Chicago Press, 1922.

Brower, Charles D. *King of the Arctic: A Lifetime of Adventure in the Far North.* London: Robert Hale, 1958.

Buckner, Phillip, ed. *Canada and the British Empire.* Oxford: Oxford University Press, 2008.

Buss, Helen M. *Undelivered Letters to Hudson's Bay Company Men on the Northwest Coast.* Vancouver: University of British Columbia Press, 2003.

Cairnes, D. D. *The Yukon-Alaska International Boundary, between Yukon and Porcupine Rivers.* Department of Mines: Geological Survey, memoir 67, no. 49 Ottawa: Government Printing Office, 1914.

Cameron, Emilie. "Copper Stories: Imaginative Geographies and Material Orderings of the Central Canadian Arctic." In *Rethinking the Great White North: Race, Nature, and the Historical Geographies of Whiteness in Canada,* edited by A. Baldwin, L. Cameron, and A. Kobayashi, 169–90. Vancouver: University of British Columbia Press, 2011.

———. *Far Off Metal River: Inuit Lands, Settler Stories, and the Making of the Contemporary Arctic.* Vancouver: University of British Columbia Press, 2015.

———. "Securing Indigenous Politics: A Critique of the Vulnerability and Adaptation Approach to the Human Dimensions of Climate Change in the Canadian Arctic." *Global Environmental Change* 22 (2012): 103–14.

Campbell, Robert Bruce. *In Darkest Alaska: Travels and Empire along the Inside Passage.* Philadelphia: University of Pennsylvania Press, 2007.

Carter, R. D., C. G. Mull, K. J. Bird, and R. B. Powers. "The Petroleum Geology and Hydrocarbon Potential of Naval Petroleum Reserve No. 4, North Slope, Alaska." US Department of Interior, US Geological Survey, April 1977.

Cassady, Josslyn. "State Calculations of Cultural Survival in Environmental Risk Assessment: Consequences for Alaska Natives." *Medical Anthropology Quarterly* 24, no. 4 (2010): 451–71.

Cavell, Janice. "Arctic Exploration in Canadian Print Culture, 1890–1930." *Papers of the Bibliographical Society of Canada* 44, no. 2 (2006): 7–43.

———. "The Second Frontier: The North in English-Canadian Historical Writing." *Canadian Historical Review* 83, no. 3 (2002): 364–89.

Cavell, Janice, and Jeff Noakes. *Acts of Occupation: Canada and Arctic Sovereignty, 1918–25.* Vancouver: University of British Columbia Press, 2010.

Coates, Kenneth. *Canada's Colonies: A History of the Yukon and Northwest Territories.* Toronto: James Lorimer, 1985.

Coates, Kenneth S., and William R. Morrison. "The New North in Canadian History and Historiography." *History Compass* 6, no. 2 (2008): 639–58.

———. "Soldier-Workers: The U.S. Army Corps of Engineers and the Northwest Defense Projects, 1942–1946." *Pacific Historical Review* 62, no. 3 (1993): 273–304.

———. "Writing the North: A Survey of Contemporary Canadian Writing on Northern Regions." *Essays on Canadian Writing* 59 (Fall 1996): 5–25.

Coates, Peter A. *The Trans-Alaskan Pipeline Controversy: Technology, Conservation, and the Frontier.* Anchorage: University of Alaska Press, 1993.

Collignon, Béatrice. "Inuit Place Names and Sense of Place." In *Critical Inuit Studies—An Anthology of Contemporary Arctic Ethnography*, edited by Pamela Stern and Lisa Stevenson, 187–205. Lincoln: University of Nebraska Press, 2006.

Collins, Henry. "Diamond Jenness: Arctic Archaeology." *Beaver* (Autumn 1967): 78–79.

Cooke, A., and C. Holland. *The Exploration of Northern Canada, 500 to 1920: A Chronology*. Toronto: Arctic History Press, 1978.

Cronon, William. "A Place for Stories: Nature, History, and Narrative." *Journal of American History* 78, no. 4 (1992): 1347–76.

Dathan, Wendy. *The Reindeer Botanist: Alf Erling Porsild, 1901–1977*. Calgary: University of Calgary Press, 2012.

De Sainville, Edouard. "Journey to the Mouth of the Mackenzie." *Geographical Journal* 13, no. 4 (1899): 435.

———. "Voyage a l'embouchure de la Riviere Mackenzie, 1889–1894." *Bulletin de la Société Géographie* 19 (1898): 291–307.

Diubaldo, Richard. *Stefansson and the Canadian Arctic*. Montreal: McGill-Queen's University Press, 1978.

Dolin, Eric Jay. *Leviathan: The History of Whaling in America*. New York: W. W. Norton, 2007.

Dorsey, Kurkpatrick. *The Dawn of Conservation Diplomacy: U.S.-Canadian Wildlife Protection Treaties in the Progressive Era*. Seattle: University of Washington Press, 1998.

Dyer, Gwynne. *Climate Wars: The Fight for Survival as the World Overheats*. Oneworld, 2011.

Easterbrook, W. T., and M. H. Watkins. "Introduction," and "Part 1: The Staple Approach." In *Approaches to Canadian Economic History*. Ottawa: Carleton University Press, 1984.

Emmerson, Charles. *The Future History of the Arctic*. New York: PublicAffairs, 2010.

Farish, Matthew. "Creating Cold War Environments: The Laboratories of American Globalism." In *Environmental Histories of the Cold War*, edited by J. R. McNeill and Corinna R. Unger, 51–84. New York: Cambridge University Press, 2010.

———. "Frontier Engineering: From the Globe to the Body in the Cold War Arctic." *Canadian Geographer* 50, no. 2 (2006): 177–96.

———. "Review of Entangled Geographies: Empire and Technopolitics in the Global Cold War." *Isis* 103, no. 4 (2012): 805–6.

Farish, Matthew, and P. Whitney Lackenbauer. "High Modernism in the Arctic: Planning Frobisher Bay and Inuvik." *Journal of Historical Geography* 35 (2009): 517–44.

Fienup-Riordan, Ann. *Freeze Frame: Alaska Eskimos in the Movies*. Seattle: University of Washington Press, 2003.

Finnie, Richard. *Canada Moves North*. New York: Hurst and Blackett, 1942.

———. *CANOL*. San Francisco: Taylor and Taylor, 1945.

Fitzhugh, William. "'Of No Ordinary Importance': Reversing Polarities in Smithsonian Arctic Studies." In *Smithsonian at the Poles: Contributions to International Polar Year Science*, edited by Igor Krupnik, Michael A. Lang, and Scott E. Miller, 61–77. Washington, DC: Smithsonian Institution Scholarly Press, 2009.

Ford, J. A. "Eskimo Prehistory in the Vicinity of Point Barrow, Alaska." *Anthropological Papers of the American Museum of Natural History* 47, part 1 (1959).

Fumoleau, Rene. *As Long as This Land Shall Last*. 2nd ed. Calgary: University of Calgary Press, 2004.

Funk, McKenzie. "Cold Rush: The Coming Fight for the Melting North." *Harper's Magazine* (September 2007): 45–55.

Godsell, Phillip. *Red Hunters of the Snows: An Account of Thirty Years' Experience with the Primitive Indian and Eskimo Tribes of the Canadian North-West and Arctic Coast, with a Brief History of the Early Contact between White Fur Traders and the Aborigines*. London: National Museum of Canada, 1938.

Goldstein, Daniel. "'Yours for Science': The Smithsonian Institution's Correspondents and the Shape of Scientific Community in Nineteenth-Century America." *Isis* 85, no. 4 (1994): 573–99.

Grace, Sherrill E. *Canada and the Idea of North*. Montreal: McGill-Queen's University Press, 2001.

Graham, Amanda. "Reflections on Contemporary Northern Canadian History." *Essays on Canadian Writing* 59 (Fall 1996): 182–200.

Grant, Shelagh D. "Northern Nationalists: Visions of 'A New North,' 1940–1950." In *For Purposes of Dominion: Essays in Honour of Morris Zaslow*, edited by Kenneth S. Coates and William R. Morrison, 47–70. North York, ON: Captus Press, 1989.

Greeley, A. W. *The Polar Regions in the Twentieth Century: Their Discovery and Industrial Evolution*. Boston: Little, Brown, 1928.

Green, Lewis. *The Boundary Hunters: Surveying the 141st Meridian and the Alaska Panhandle*. Vancouver: University of British Columbia Press, 1982.

Gruening, Ernest. *The State of Alaska*. New York: Random House, 1954.

———, ed. *An Alaskan Reader, 1867–1967*. New York: Meredith Press, 1966.

Hacking, Ian. *Representing and Intervening: Introductory Topics in the Philosophy of Natural Science*. Cambridge: Cambridge University Press, 1983.

Hall, Sam. *The Fourth World: The Heritage of the Arctic and Its Destruction*. London: Bodley Head, 1987.

Harris, R. Cole. *The Resettlement of British Columbia: Essays on Colonialism and Geographical Change*. Vancouver: University of British Columbia Press, 1997.

Harrison, Alfred H. *In Search of a Polar Continent, 1905–1907*. London: Edward Arnold, 1908.

Haycox, Stephen. *Alaska: An American Colony*. Seattle: University of Washington Press, 2006.

———. *Frigid Embrace: Politics, Economics, and Environment in Alaska*. Corvallis: Oregon State University Press, 2002.

Hays, Samuel P. *Conservation and the Gospel of Efficiency: The Progressive Conservation Movement, 1890–1920*. Cambridge, MA: Harvard University Press, 1959.

Hill, Dick. *Inuvik: A History, 1958–2008: The Planning, Construction, and Growth of an Arctic Community*. Victoria, BC: Trafford Publishing, 2008.

Hinckley, Ted C. *The Americanization of Alaska, 1867–1897*. Palo Alto, CA: Pacific Books, 1972.

Hinsley, Curtis. *Savages and Scientists: The Smithsonian Institution and the Development of American Anthropology, 1846–1910*. Washington, DC: Smithsonian Institution Press, 1981.

Ingram, Rob, and Helene Dobrowolsky. *Waves upon the Shore: A Historical Profile of Herschel Island*. Prepared for Heritage Branch, Department of Tourism, Government of Yukon Territory, September 1989.

Jenness, Diamond. "The Copper Eskimos." *Geographical Review* 4, no. 2 (1917): 81–91.

———. *Dawn in Arctic Alaska*. Chicago: University of Chicago Press, 1957.

———. "The Eskimos of Northern Alaska: A Study in the Effect of Civilization." *Geographical Review* 5 (1918): 89–101.

———. "Note on Cadzow's 'Native Copper Objects of the Copper Eskimo.'" *American Anthropologist*, n.s., 23, no. 2 (1921): 235–36.

———. *People of the Twilight*. New York: Macmillan, 1928.

Jenness, Stuart E. *Arctic Odyssey: The Diary of Diamond Jenness, Ethnologist with the Canadian Arctic Expedition in Northern Alaska and Canada, 1913–1916*. Hull, QC: Canadian Museum of Civilization, 1991.

———. *Stefansson, Dr. Anderson, and the Canadian Arctic Expedition, 1913–1918*. Gatineau, QC: Canadian Museum of Civilization, 2011.

Jørgensen, Dolly, and Sverker Sörlin, eds. *Northscapes: History, Technology, and the Making of Northern Environments*. Vancouver: University of British Columbia Press, 2013.

Kirsch, Scott. *Proving Grounds: Project Plowshare and the Unrealized Dream of Nuclear Earthmoving*. New Brunswick, NJ: Rutgers University Press, 2005.

Kitto, F. H. "The Northwest Territories, 1930." Department of the Interior, Northwest Territories and Yukon Branch, 1933.

Kohler, Robert E. *All Creatures: Naturalists, Collectors, and Biodiversity, 1850-1950*. Princeton, NJ: Princeton University Press, 2006.

———. "Finders, Keepers: Collecting Sciences and Collecting Practice." *History of Science* 45 (2007): 428-53.

———. "History of Field Science: Trends and Prospects." In *Knowing Global Environments: New Historical Perspectives on the Field Sciences*, edited by Jeremy Vetter, 212-40. New Brunswick, NJ: Rutgers University Press, 2011.

———. *Landscapes and Labscapes: Exploring the Lab-Field Border in Biology*. Chicago: University of Chicago Press, 2002.

Krupnik, Igor. "From Boas to Burch: One Hundred Years of 'Eskimology,' 1880-1980." Presentation, Inuit Studies Conference, Washington, DC, October 25, 2012.

Kulchyski, Peter. "Anthropology in the Service of the State: Diamond Jenness and Canadian Indian Policy." *Journal of Canadian Studies* 28, no. 2 (1993): 21-50.

Kulchyski, Peter, and Frank Tester. *Kiumajut (Talking Back): Game Management and Inuit Rights, 1900-1970*. Vancouver: University of British Columbia Press, 2007.

Lackenbauer, P. Whitney, and Matthew Farish. "The Cold War on Canadian Soil: Militarizing a Northern Environment." *Environmental History* 12 (October 2007): 920-50.

Lands, Parks, and Forests Branch. *Canada's Reindeer*. Ottawa: Northwest Territories Administration, 1940.

Le Bourdais, D. *Stefansson: Ambassador of the North*. Montreal: Harvest House, 1963.

Leffingwell, Ernest de Koven. "The Canning River Region: Northern Alaska." Professional Paper 109. US Geological Survey, 1919.

———. "My Polar Explorations, 1901-1914." *Explorers Journal* 39, no. 3 (1961): 2-14.

Leopold, A. Starker, and F. Fraser Darling. *Wildlife in Alaska: An Ecological Reconnaissance*. New York: Ronald Press, 1953.

Levere, Trevor H. *Science and the Canadian Arctic: A Century of Exploration, 1818-1918*. New York: Cambridge University Press, 1993.

———. "Vilhjalmur Stefansson, the Continental Shelf, and a New Arctic Continent." *British Journal for the History of Science* 21, no. 2 (1988): 233-47.

Libbey, David, and William Schneider. "Fur Trapping on Alaska's North Slope." In *Le Castor Fait Tout: Selected Papers of the Fifth North American Fur Trade Conference*, edited by Bruce G. Trigger, Toby Morantz, and Louise Dechêne, 335-58. Montreal: Lake St. Louis Historical Society, 1987.

Lindsay, Debra. *Science in the Subarctic: Trappers, Traders, and the Smithsonian Institution*. Washington, DC: Smithsonian Institution Press, 1993.

Lomen, Carl J. *Fifty Years in Alaska*. New York: David McKay, 1954.

Luedecke, Cornelia. "The First International Polar Year (1882-1883): A Big Science Experiment with Small Science Equipment." *Proceedings of the International Commission on History of Meteorology* 1, no. 1 (2004): 56.

Lyons, Natasha. "Inuvialuit Rising: The Evolution of Inuvialuit Identity in the Modern Era." *Alaska Journal of Anthropology* 7, no. 2 (2009): 63-79.

Lyons, Natasha, with a contribution by Sandra Jezik. "Early 20th Century Lifeways in Qainiuqvik (Clarence Lagoon): Report on the 2003 Archaeological Investigations at Site 76Y." Cultural Resource Services, Parks Canada, 2004.

MacFarlane, Roderick. "On an Expedition down the Begh-ula or Anderson River." *Canadian Record of Science* 4 (1891): 28-53.

Mackay, J. R., W. H. Mathews, and R. S. MacNeish. "Geology of the Engigstciak Archaeological Site, Yukon Territory." *Arctic* 14, no. 1 (1961): 25-52.

Mair, Charles, and Roderick MacFarlane. *Through the Mackenzie Basin*. Toronto: William Briggs, 1908.

Martin-Nielsen, Janet. "The Other Cold War: The United States and Greenland's Ice Sheet Environment, 1948-1966." *Journal of Historical Geography* 38 (2012): 69-80.

McCannon, John. *A History of the Arctic: Nature, Exploration, and Exploitation.* London: Reaktion Books, 2012.

McGhee, Robert. *Beluga Hunters: An Archaeological Reconstruction of the History and Culture of the Mackenzie Delta Kittegaryumiut.* Saint Johns: Institute of Social and Economic Research, Memorial University of Newfoundland, 1974.

McNeill, J. R. *Something New under the Sun: An Environmental History of the Twentieth Century World.* New York: Norton, 2000.

——, and Corinna R. Unger, eds. *Environmental Histories of the Cold War.* New York: Cambridge University Press, 2010.

McNeish, R. S. "Archaeological Reconnaissance of the Delta of the Mackenzie River and Yukon Coast." *National Museum of Canada Bulletin* 142 (1956).

Mikkelsen, Ejnar. *Conquering the Arctic Ice.* Philadelphia: George W. Jacobs, 1909.

Morrison, David. "An Archaeological Perspective on Neoeskimo Economies." In *Threads of Arctic Prehistory: Papers in Honour of William E. Taylor, Jr.*, edited by David Morrison and Jean-Luc Pilon, 311-20. Ottawa: Canadian Museum of Civilization, 1994.

Morrison, William R. *Showing the Flag: The Mounted Police and Canadian Sovereignty in the North, 1894-1925.* Vancouver: University of British Columbia Press, 1985.

Morse, Kathryn. *The Nature of Gold: An Environmental History of the Klondike Gold Rush.* Seattle: University of Washington Press, 2003.

Mulvihill, Peter R., Douglas C. Baker, and William R. Morrison. "A Conceptual Framework for Environmental History in Canada's North." *Environmental History* 6, no. 4 (2001): 611-26.

Nadasdy, Paul. "The Politics of TEK: Power and the 'Integration' of Knowledge." *Arctic Anthropology* 36, nos. 1-2 (1999): 1-18.

Naske, Claus-M., and Herman E. Slotnick. *Alaska: A History.* Norman: University of Oklahoma Press, 2011.

Neufeld, David. "Commemorating the Cold War in Canada: Considering the DEW Line." *Public Historian* 20, no. 1 (1998): 9-19.

——. "Distant Early Warning (DEW) Line: A Preliminary Assessment of Its Role and Effects upon Northern Canada." Revised for the Arctic Institute of North America, May 2002.

Neufeld, Peter Lorenz. "De Sainville: Forgotten Mackenzie Mapper." *North* (1981): 55-57.

"The New North." *Nature* 478 (October 13, 2011): 172-74.

North, Dick. *Arctic Exodus: The Last Great Trail Drive.* Guilford, CT: Lyons Press, 1991.

North Slope Borough Contract Staff. *Native Livelihood and Dependence: A Study of Land Use Values through Time.* Anchorage: US Department of the Interior, National Petroleum Reserve in Alaska, 1979.

Numbers, Ronald, and Charles Rosenberg, eds. *The Scientific Enterprise in America: Readings from Isis.* Chicago: University of Chicago Press, 1996.

Nuttall, Mark. *Pipeline Dreams: People, Environment, and the Arctic Energy Frontier.* International Work Group for Indigenous Affairs, 2010.

Olson, Dean F. *Alaska Reindeer Herdsmen: A Study of Native Management in Transition.* Fairbanks, AK: Institute of Social, Economic, and Government Research, 1969.

O'Neill, Dan. *The Firecracker Boys: H-Bombs, Inupiat Eskimos, and the Roots of the Environmental Movement.* New York: Basic Books, 2007.

Owram, Doug. *Promise of Eden: The Canadian Expansionist Movement and the Idea of the West.* Toronto: University of Toronto Press, 1992.

Palmer, Lawrence J. "The Alaska Tundra and Its Use by Reindeer." US Department of the Interior, Office of Indian Affairs, 1945.

——. *Progress of Reindeer Grazing Investigations in Alaska.* Bulletin no. 1423, October 1926. Washington, DC: US Department of Agriculture, 1927.

Palmer, L. J., and Seymour Hadwen. Reindeer in Alaska. Bulletin no. 1089, September 22, 1922. Washington, DC: US Department of Agriculture, 1926.

Pálsson, Gísli, ed. *Writing on Ice: The Ethnographic Notebooks of Vilhjalmur Stefansson.* Hanover, NH: University Press of New England, 2001.

Pimlott, Douglas. "Offshore Drilling in the Beaufort Sea: A Report to the Committee for Original People's Entitlement (COPE)." Prepared by Canadian Arctic Resources Committee. Inuvik, NT, 1974.

Piper, Liza. *The Industrial Transformation of Subarctic Canada.* Vancouver: University of British Columbia Press, 2009.

Piper, Liza, and John Sandlos. "A Broken Frontier: Ecological Imperialism in the Canadian North." *Environmental History* 12 (2007): 759–95.

Porsild, Alf Erling. *Reindeer Grazing in Northwest Canada: Report of an Investigation of Pastoral Possibilities in the Area from the Alaska-Yukon Boundary to Coppermine River.* Ottawa: F. A. Acland, 1929.

Porsild, Erling. "The Reindeer Industry and the Canadian Eskimo." *Geographical Journal* 88, no. 1 (1936): 1–16.

Powell, Richard C. "'The Rigours of an Arctic Experiment': The Precarious Authority of Field Practices in the Canadian High Arctic, 1958–1970." *Environment and Planning A* 39 (2007): 1794–811.

Pratt, Mary Louise. *Imperial Eyes: Travel Writing and Transculturation.* New York: Routledge, 1992.

Puiguitkaat (The 1978 Elder's Conference). Transcription and translation by Leona Okakok, edited by Gary Kean. Barrow, AK: North Slope Borough Commission on History and Culture, 1981.

Rabbit, Mary C. *Minerals, Lands, and Geology for the Common Defence and General Welfare: A History of Public Lands, Federal Science and Mapping Policy, and Development of Mineral Resources in the United States.* Vol. 3, 1904–1939. Washington, DC: US Geological Survey, 1986.

Raibmon, Paige. *Authentic Indians: Episodes of Encounter from the Late Nineteenth-Century Northwest Coast.* Durham, NC: Duke University Press, 2005.

Ray, Arthur. "Recent Trends in Northern Historiography." *Essays on Canadian Writing* 59 (Fall 1996): 211–29.

Reiss, Bob. *The Eskimo and the Oil Man: The Battle at the Top of the World for America's Future.* New York: Business Plus, 2012.

Riffenbaugh, Beau. *The Myth of the Explorer: The Press, Sensationalism, and Geographical Discovery.* New York: Belhaven Press, 1993.

Robertson, Gordon. *Memoirs of a Very Civil Servant: Mackenzie King to Pierre Trudeau.* Toronto: University of Toronto Press, 2000.

Robinson, Michael. *The Coldest Crucible: Arctic Exploration and American Culture.* Chicago: University of Chicago Press, 2006.

Russell, Frank. *Explorations in the Far North: Being the Report of an Expedition under the Auspices of the University of Iowa during the Years 1892, '93, and '94.* Iowa City: University of Iowa, 1898.

Sabin, Paul. "Voices from the Hydrocarbon Frontier: Canada's Mackenzie Valley Pipeline Inquiry (1974–1977)." *Environmental History Review* 19, no. 1 (1995): 17–48.

Safier, Neil. "Global Knowledge on the Move: Itineraries, Amerindian Narratives, and Deep Histories of Science." *Isis* 101 (2010): 133–45.

Salvatore, Ricardo D. "The Enterprise of Knowledge: Representational Machines of Informal Empire." In *Close Encounters of Imperialism: Writing the Cultural History of U.S.-Latin American Relations,* edited by Gilbert M. Joseph, Catherine C. Legrand, and Ricardo D. Salvatore. Durham, NC: Duke University Press, 1998.

Sandlos, John. *Hunters at the Margin: Native People and Wildlife Conservation in the Northwest Territories.* Vancouver: University of British Columbia Press, 2007.

———. "Landscaping Desire: Poetics, Politics in the Early Biological Surveys of the Canadian North." *Space and Culture* 6 (2003): 396–409.

————. "Where the Scientists Roam: Ecology, Management, and Bison in Northern Canada." In *Canadian Environmental History: Essential Readings*, edited by David Freeland Duke, 333-60. Toronto: Canadian Scholars' Press, 2006.

Scott, James C. *Seeing Like a State: How Certain Schemes to Improve the Human Condition Have Failed.* New Haven, CT: Yale University Press, 1999.

Scotter, George W. "Reindeer Husbandry as a Land Use in Northwestern Canada." *Proceedings of the Productivity and Conservation in Northern Circumpolar Lands Conference.* Edited by W. A. Fuller and P. G. Kevan. IUCN, n.s., no. 16 (1969): 159-69.

Seguin, Gilles. "Reindeer for the Inuit: The Canadian Reindeer Project, 1929-1960." *Muskox* 38 (1991): 6-26.

Shelesnyak, M. C. "The History of the Arctic Research Laboratory, Point Barrow, Alaska." *Arctic* 1, no. 2 (1948): 97-106.

Sherwood, Morgan. *Big Game in Alaska.* New Haven, CT: Yale University Press, 1981.

————. *Exploration of Alaska, 1865-1900.* New Haven, CT: Yale University Press, 1965.

Smith, Laurence C. *The World in 2050: Four Forces Shaping Civilization's Northern Future.* New York: Dutton, 2010.

Smith, Neil. *American Empire: Roosevelt's Geographer and the Prelude to Globalization.* Berkeley: University of California Press, 2003.

Sonnenfeld, Joseph. "An Arctic Reindeer Industry: Growth and Decline." *Geographical Review* 49, no. 1 (1959): 76-94.

Sörlin, Sverker. "Rituals and Resources of Natural History: The North and the Arctic in Swedish Scientific Nationalism." In *Narrating the Arctic: A Cultural History of Nordic Scientific Practices,* edited by Michael Bravo and Sverker Sörlin, 73-122. Canton, MA: Science History Publications, 2002.

————. *Science, Geopolitics, and Culture in the Polar Region: Norden beyond Borders.* Burlington, VT: Ashgate, 2013.

Stager, John K. "Fort Anderson: The First Post for Trade in the Western Arctic." *Geographical Bulletin* 9, no. 1 (1967): 46-48.

Stefansson, Vilhjalmur. "The Arctic Headlines in Fact and Fable." *Foreign Policy Association* 51 (January 1945).

————. "Eskimo Trade Jargon of Herschel Island." *American Anthropologist* 11, no. 2 (1909): 217-32.

————. *The Friendly Arctic: The Story of Five Years in the Polar Regions.* New York: Macmillan, 1921.

————. "Icelandic Beast and Bird Lore." *Journal of American Folklore* 19, no. 75 (1906): 300-308.

————. "The Icelandic Colony in Greenland." *American Anthropologist* 8, no. 2 (1906): 262-70.

————. *My Life with the Eskimo.* New York: Macmillan, 1913.

————. "My Quest in the Arctic." *Harper's Monthly Magazine* (December 1912).

————. *The Northward Course of Empire.* London: George G. Harrap, 1922.

————. "Suitability of Eskimo Methods of Winter Travel in Scientific Exploration." *Bulletin of the American Geographical Society of New York* 60, no. 1 (1908): 210-13.

————. "The Technique of Arctic Winter Travel." *Bulletin of the American Geographical Society of New York* 44, no. 1 (1912): 340-47.

Stocking, George W., Jr. "Anthropology as Kulturkampf: Science and Politics in the Career of Franz Boas." In *The Ethnographer's Magic and Other Essays in the History of Anthropology,* by George Stocking Jr., 92-113. Madison: University of Wisconsin Press, 1992.

————. "Ideas and Institutions in American Anthropology: Thoughts toward a History of the Interwar Years." In *The Ethnographer's Magic and Other Essays in the History of Anthropology,* by George Stocking Jr., 114-77. Madison: University of Wisconsin Press, 1992.

Stommel, Henry. *Lost Islands: The Story of Islands That Have Vanished from Nautical Charts.* Vancouver: University of British Columbia Press, 1984.

Stradling, David, ed. *Conservation in the Progressive Era: Classic Texts.* Seattle: University of Washington Press, 2004.

Stuck, Hudson. *A Winter Circuit of Our Arctic Coast: A Narrative of a Journey with Dog-Sleds around the Entire Arctic Coast of Alaska.* New York: Charles Scribner's Sons, 1920.

Sutter, Paul. "What Can U.S. Environmental Historians Learn from Non-U.S. Environmental Historiography?" *Environmental History* 8, no. 1 (2003): 109-29.

Swazye, Nansi. *Canadian Portraits: Jenness, Barbeau, Wintemberg; The Man Hunters.* Toronto: Clarke, Irwin, 1960.

Taber, Stephen. "Perennially Frozen Ground in Alaska: Its Origin and History." *Geological Society of America Bulletin* 54 (1943): 1433-548.

Taliafero, John. *In a Far Country: The True Story of a Mission, a Marriage, a Murder, and the Remarkable Reindeer Rescue of 1898.* New York: PublicAffairs, 2006.

Taylor, John Leonard. *Canadian Indian Policy during the Inter-War Years, 1918-1939.* Ottawa: Department of Indian Affairs and Northern Development, 1984.

Taylor, Joseph E., III. "The Many Lives of the New West." *Western Historical Quarterly* 35, no. 2 (2004): 141-65.

Tester, Frank, and Peter Kulchyski. *Tammarniit (Mistakes): Inuit Relocation in the Eastern Arctic, 1939-1963.* Vancouver: University of British Columbia Press, 1994.

University of California. "George Collins: The Art and Politics of Park Planning and Preservation: An Interview." Regional Oral History Office, Bancroft Library, 1980, 215-25.

Usher, Peter J. "Eskimo Land Use and Occupancy in the Western Arctic." Prepared for Milton Freeman Research. Hamilton, Ontario, 1974.

———. *Fur Trade Posts of the Northwest Territories, 1870-1970.* Ottawa: Northern Science Research Group, Department of Indian Affairs and Northern Development, 1971.

Valentine, V. F., and J. R. Lotz. "Northern Co-Ordination and Research Centre of the Canadian Department of Northern Affairs and National Resources." *Polar Record* 11, no. 73 (1963): 419-22.

Willis, Roxanne. *Alaska's Place in the West: From the Last Frontier to the Last Great Wilderness.* Lawrence: University Press of Kansas, 2011.

———. "A New Game in the North: Native Reindeer Herding, 1890-1940." *Western Historical Quarterly* 37 (Autumn 2006): 277-301.

Wissler, Clark. *An Introduction to Social Anthropology.* New York: Henry Holt, 1929.

———. "'Turning Kogmollik' for Science." *American Museum Journal* 10, no. 7 (1910): 212-20.

Wright, Edmund A. *CRREL's First 25 Years, 1961-1986.* U.S. Army Cold Regions Research and Engineering Laboratory, June 1986.

Young, Christian C. "Defining the Range: The Development of Carrying Capacity in Management Practice." *Journal of the History of Biology* 31 (1998): 61-83.

Zaslow, Morris. *The Northward Expansion of Canada, 1914-1967.* Toronto: McClelland and Stewart, 1988.

———. *The Opening of the Canadian North, 1870-1914.* Toronto: McClelland and Stewart, 1971.

———. *Reading the Rocks: The Story of the Geological Survey of Canada, 1842-1972.* Toronto: Macmillan, 1975.

Zeller, Suzanne. *Inventing Canada: Early Victorian Science and the Idea of a Transcontinental Nation.* Montreal: McGill-Queen's University Press, 2009.

Index

Page numbers in italic indicate figures and maps.